# TECHNIQUES
## OF TRAINING IN
# VALUE
# ENGINEERING
## TRAINERS MANUAL

I0483383

## R.G. CHAUDHARI

# Notion Press

Old No. 38, New No. 6
McNichols Road, Chetpet
Chennai - 600 031

First Published by Notion Press 2018
Copyright © R.G. Chaudhari 2018
All Rights Reserved.

ISBN 978-1-64249-127-2

# Contents

# FOREWORD

It is with great pleasure that I respond to your invitation to add a few words to your **exceptional VE Manual**.

**I know of no other VE publication that has so successfully assembled so many relevant aspects of the VE story. This manual is bound to become a classic as soon as it is published.**

**The careful detailed development of all important aspects of the VE story reflects the monumental scope of the publication.**

During the last seven years I have spent a month each year giving a series of VE workshop/seminars in India under the auspices of the Indo-American Society of Bombay. I have noted a few aspects of VE as it is practiced in India that I believe are worth recording here.

1. Because VE is primarily an intellectual exercise which makes the most cost-effective use of materials, it is a technique especially useful in pre-industrial societies. This is particularly true when there is a large educated un – (or under) employed sector of the population.

2. During the seven years I have been talking of VE here, the general understanding of the discipline has gone from emphasis on its cost-cutting liabilities to understanding its use to improve quality, until today when it is more widely appreciated in "updating" designs which were sufficient when they were first deployed.

This wider understanding expressed above has come from a relatively small group of publications. **Now we have this large, almost overwhelming manual which certainly will make its positive contribution to the advancement of the state of art in India. …….. and wherever it may be read.**

**Finally, special note must be made of the many clear insights in the description of various phases of the VE job plan in this manual.**

*(Signed)*

Harold C. Tufty, CVS F. SAVE                                                                                    6th March 1987

Editor, Publisher

Value Engineering & Management Digest

1199 National press building

Washington D.C. 20045

# Prof. Harold  G.Tufty

## (Writer of  FOREWORD  to this Manual)

He is presently (1987) the President of Tufty & Associates, a consulting firm in the areas of  Value Engineering, Management, Design and Construction. He is the Publisher of 'Value Engineering Digest' and has done extensive consultation work in Value Engineering  for construction management  firms, universities.

Architect/engineering companies and US  Federal Government Agencies. He was a team leader of the VE Team which performed a successful study of US Senators office (Clerk–Lowa).

Prof. Tufty  is the past President of the National Capital chapter of  the Society Of American Value Engineers (SAVE) and is a former National Vice President of SAVE. He has been actively engaged in Value Engineering since 1966 and is currently a Board Member of the National Society and the local Chapter. He is also President of the recently formed Value Foundation.

# PREFACE

The manual is designed for the use of practicing and aspiring value specialists desiring to take up training in Value Engineering/Value Management techniques as part of their consultancy service or full time career. The contents of the manual are based on proven and time-tested training techniques including TWI methodology. They have already proven useful to me in my job over two and a half decades in engineering industry as a trainer, as an industrial engineer, and on the shop floor. It is my earnest hope that this training methodology may prove as useful to you, particularly in conducting in-company VE workshops for working level engineers, executives, technicians and supervisors.

VE/VM is not only a most cost effective Management tool but also a fascinating field of study. Fascinating–because it revolves round the unique faculty of human mind, namely, creative imagination and offers unlimited opportunities for learning and developing the power of thinking in all fields of human activities.

Value Engineering is the one and only one method "to balance infinite wants with infinite resources," but it has not yet received the attention to the extent it so deserves even from the professionally managed establishments in our country. One plausible explanation is non-availability of VE practitioners in sufficient numbers on one hand, and inadequate training aids on the other hand. I sincerely hope that this manual may to a great extent, lessen the inadequacies in training materials and fulfil the need of trained manpower in the use of VE/VM techniques.

I have drawn freely from the inputs provided by Indo-American Society, Bombay (Mumbai), Institute of Creativity Development, Poona (Pune) and National Productivity Council, New Delhi, during the programmes conducted by them.

I also relied on numerous case studies and papers presented by my co- participants in VE programs conducted by IIM (Bangaluru), NPC (Bangaluru & Kalkata) and International Seminar on VE held at Jamshedpur in June 1980. I am indebted to these institutions of which inputs have helped me a lot in writing this Manual.

I acknowledge my gratitude to the Management of Bharat Heavy Electrical ltd. for giving me many, many opportunities to participate in and to conduct numerable executive and supervisory development programmes and to put into practice what was taught in the classrooms *My special thanks are due to Mr. GSR Subramanyam, my PA, who had painstakingly typed the MS of this manual.*

My thanks are also due to my erstwhile colleagues in Training & Development, Industrial Engineering and/Productivity Services, Prototype Development Centre and to many more trainees and participants in the programmes and seminars, who have inspired me directly and indirectly in undertaking and completing this work.

Hyderabad-India                                                                                                              R.G. Chaudhari

# WHY VALUE ENGINEERING?

Value Analysis/Value Engineering (VA/VE) is a post-world war-II innovation, inspired by Harry Erlicher and systematised by L. D. Miles, both of GEC, USA, in 1947. Within few years of its formulation, the VA/VE was widely used in GEC, not only in purchasing where it was born, but also in many other functional areas, yielding richest dividends in Product development and Cost Reduction activities. Such was the success of VA/VE in GEC that it soon caught the attention of and Spread to, other enterprises in USA within a short period of 5 to 6 years. In the first two decades of its application, countries like USA, Japan and Germany have saved millions of dollars through VE techniques. USA Navy Bureau of Ships, General Dynamics, USA Army, Corps of Engineers and other government and private agencies have reported returns of 20:1 and higher, on investment. More important than this, the VE has proven that it is a universal and powerful management tool in revealing unnecessary costs and in providing quality goods and services at the lowest cost without compromising on performance and reliability.

Inspite of its being most potent and most effective in counter- acting shortages, adverse effects of inflation, energy crises and stiff market competition, the VE remains a somewhat neglected discipline even in large industries, Defence Services, Government & Quasi-Government establishments. One possible reason for this neglect may be lack of awareness of its immense potentialities or lack of knowledge or confidence in its application. To my mind an ideal approach lies in building up confidence through systematic training of working level executives and constant review of their achievement and all on-going activity.

The value specialists being few in number, many organizations might find it difficult to get their services at frequent intervals and at affordable cost. Also, the small and medium size enterprises may find it un-economical to retail a value specialist on a full time basis as a trainer. The compromise solution in my opinion is to organize a continuing programmes of training by utilizing services of one or two in-company executives after exposing them to the use of training techniques and, backing up their efforts with due encouragement. Magnitude of operation in medium and large industries would justify such continuing programmes and investment. Small industries can pool their resources and employ the services of the consultant to exploit the benefits of VE to their units, Such on-going scheme needs structured lesson plans with a number of live cases with varied fields of economic activity, and this manual precisely aims at meeting that need.

The training methodology followed in the manual encourages full involvement of the participants, right from reaching agreement on principles to taking decisions on selected alternatives and implementation.

I hope the manual will act as a boon to the practitioners and aspiring value specialists to become proficient trainers in Value Management. The value trained people intuitively survey and probe their own activities, and the activities of others, and apply or help in applying the VE techniques, as a force of habit, resulting in rich dividend to their organization.

Hyderabad-India

R.G. Chaudhari, CVS

20 June,1988

Reviewed: Jan 2017

# 1

# VALUE ENGINEERING WORKSHOP

**GENERAL GUIDELINES:**

1. FACILITIES & ARRANGEMENT

2. HINTS FOR THE TRAINER

3. MECHANICS OF PRESENTATION

4. USING THE MANUAL

5. SESSION-WISE THINGS YOU NEED

6. SUGGESTED TIME SCHEDULE

# 1. FACILITIES AND ARRANGEMENTS

1.01 First requirement is a hall large enough to accommodate about 20 members seated in conference style and a row of about 12 extra chairs along the walls. If the hall is not large enough to accommodate 4 or 5 syndicates each consisting of 4 or 5 members, at least one room adjacent to the hall should be provided for syndicate work.

1.02 The rooms should be well ventilated and free from exterior noises. The temperature and humidity in the room are important; excessive heat or cold distracts concentration.

1.03 The sitting arrangement should be such that the members should be able to see each other's faces, as in conference. Table arranged in 'U' or 'V' formation can meet this requirement.

1.04 A large sized chalk board not less than 4' × 5' or a rolling type of chalk board, ample supplies of white and coloured chalks and good quality of at least two dusters are indispensible.

1.05 If smoking is permitted ash-trays should be provided.

1.06 There are a number of good films available on VE/VM and Creativity with run time of about half an hour each. Efforts should be made to screen at least two such films during the workshop.

1.07 The suggested flip charts and transparencies for overhead projections should be prepared in advance, and should look impressive – with a touch of professional artist. (They should not look like pages from a scribbling pad as it usually happens). You may use electronic gadgets.

1.08 If flip charts are used, a good quality easel or flip-chart board should be provided. A slide projector, film strip/film projector, a tape recorder, overhead projector, and any other audio/video aids like DVD Player that are planned to be used, should be arranged and tested at least one day in advance and left in their room in the well appointed places with readiness for operation at the flick of a switch. (See appendix-1: Specimens of flip charts/transparencies for overhead projector or slides).

1.09 It should be insured that the inputs to the demonstration equipment like films, tapes and slides are in usable condition and in proper sequence.

1.10 The hand-outs planned to be distributed during different sessions should be stored preferably one day in advance in a side board kept in the class room.

1.11 Name Boards/Cards of the participants, the trainer, the chief guest, the Value Coordinator and other management representatives who are likely to address the participants at different points of time, should be prepared well before the commencement of the VE workshop, taking utmost care to check the correctness of the initials and spellings of the names on the boards/cards. The boards should be kept on the tables at least 15 minutes before the start of the 1st session. Participants' name boards should be left on the top of the tables till the end of the workshop.

Name Strips of the size of 20mm × 80mm with names typed on computer and laminated, that can be clipped to shirt pockets are preferred to aforesaid boards/cards.

1.12 Exhibitions make very effective and positive impacts on the minds of observers. Every trainer should strive hard to arrange an exhibition of VA/VE projects in the vicinity of the venue of the VE workshop, using charts. models and real jobs. It is advisable to have a permanent exhibition of such projects in the organization, built up over a period of time.

1.13 Visits of other organizations where VA/VE techniques are practiced regularly have their own worth. The trainer/VE coordinator should arrange at least one such visit during the VE programme, after ascertaining the convenience of the host organization and availability of independent transport.

1.14 Arrangement to serve tea/coffee and to host one dinner preferably with spouses of the participants and guests on the last day, should be made and service timings included in the time table.

1.15 Sufficient copies of speeches to be delivered by the representatives of the management and the guests must be made available to the audience before the start of the respective addresses. (See item No. 2.11)

1.16 Above all, it should be ensured at least one day in advance that the nominated persons or their substitutes are participating in the VE workshop and will be in their seats 15 minutes before the start of the inaugural session.

1.17 It is equally important that the invitations to the guests/invitees are in their hands 2 or 3 days before the occasion. If not, arrangement should be made to deliver the invitations through a special messenger or conveyed personally by the trainer/coordinator/facilitator.

1.18 One of the time tested techniques of motivating the participants to give their best is to promote healthy competition and award prizes or rolling shield to the best team/its members. If this idea finds approval of the Management, the prizes/shields should be procured and exhibited in the hall on the inaugural day with some catchy slogans kept underneath them.

# 2. HINTS FOR THE TRAINER

2.01     Be in the room 15 to 20 minutes before time to start the session.

2.02     Check the facilities listed in the previous pages are in their places in reasonably good/operating condition.

2.03     Put the participants in relaxed mood with a friendly word as they arrive.

2.04     Never comment or give impression of disapproval, on personal attires or on late coming, of any member.

2.05     Be cheerful and enthusiastic. Your attitude and mood are the trend setter for the others. Remember the programme can be no way better than you.

2.06     **Know the manual thoroughly which can be done only by reading it constantly.**

2.07     Vary the presentation to suit the group; change the wordings and examples to suit the composition of the group and use the manual as a guide.

2.08     Watch your timing: start and finish promptly. Do not treat each session as a complete entity; if time permits, bring forward first portion of the next session; if you are short of time, defer the left–over position of the session to the next seating. But do not omit any portion at any time (except that are indicated as OPTIONAL in the manual).

2.09     Do not attempt to lecture or domineer. Any attempt to bluff or argue the participants into accepting your view, will fail. Respect their views and feelings and steer clear controversies.

2.10     Optional: To give realistic picture of "Costs," increase the monetary values given in the manual by 5 % to 10 %  every year, depending on the rate of inflation.

2.11     Follow the well-known teaching etiquettes, a few important amongst them being the followings:

      a)    Plan your board work; make full use of the space as suggested in the manual (Procedure column).

      b)    Do not over-write obliquely or haphazardly.

      c)    Hold the chalk near its end and write in a style which is natural to you but ensure that writing is legible and in bold strokes.

      d)    **Do not present your back to the audience while writing on the board.**

      e)    Talk while writing on chalk board but raise voice.

      f)    Do not pick up chalk pieces from the floor.

      g)    Do not look at only one or two members at all times but look around the whole group and speak in a well modulated tone.

h)  Avoid actions like rolling a shirt button in fingers or chewing something in mouth or rolling cigarette on the lips, removing your spectacles every now and then, shaking of legs, tapping of feet or drumming of fingers, etc. Such actions distract the attention of the participants and is irritating to the viewers.

i)  To avoid monotony and boredom, vary your tone frequently and do not talk continuously for more than 15 minutes. The manual provides necessary breaks at appropriate intervals.

Also avoid repeated use of expressions like "you see," "you know," "catch my point," etc.

2.12  Extend help in preparation of inaugural and valedictory addresses by the head of the organization/chief guest. This can be done in two ways:

a)  Preparing their scripts and submitting for their perusal and modification; and

b)  Listing out all the salient features of VE techniques and their relevance to today's situation in your organization.

The inaugural speech by the head of the organization should contain, besides importance of the workshop enough hints to the participants that the Management is seriously interested in the workshop, that the participants should ensure their attendance in all the sessions and that the techniques should be practiced not only during the Workshop but also as part of their normal duties.

2.13.  The Trainer should procure sufficient numbers of copies of the following 2 books by the same author from the Publisher, for distribution amongst the participants of the Workshop on TECHNIQUES OF TRINING IN VALUE ENGINEERING:

a) Hand Outs on VALUE ENGINEERING AND CREATIVITY.

b) GUIDELINES FOR PREPARING VALUE ENGINEERING REPORT.

---

### Important Note for the Trainer/Facilitator

MS of this manual was prepared in 1987 taking the live case studies, mostly conducted by the author.

Figures in Rupees (INR) given in session–II were valid in 1987. If an attempt is made to match them with the present costs, by multiplying, say by 8 or 10, there will be a cascading effect on values in tables in Annexure II of the Manual.

It is the percentages of values that are important and they will not change by multiplying them by 8 or 10. Therefore, it is better to leave them as they are.

You may substitute the Case Studies given in this manual by your own examples.

# 3. MECHANICS OF PRESENTATION

## 3.01 Discussion Leading

The manual relies heavily on Group-discussion, and the success of this technique lies in getting full participation from the members, and maintaining a correct line of discussion. This calls for thorough preparation on the part of discussion leader i.e. you. It is therefore, essential that you should know the manual thoroughly, give previous thought to the subject matter, and understand the main and subsidiary questions to be posed to reach the desired conclusions. It is equally important to have a large fund of practical examples, particularly from the industries or professions to which the participants belong to illustrate the points under discussion.

The audio-visual aids, even the chalk board itself, and demonstrations from you and the participants, are the aids to the discussion and you should be able to choose and use the best of them at appropriate time.

## 3.02 Opening the Discussions

You should open the discussion by giving a brief introduction to the subject, stating a few facts and quoting well-known opinions. These statements should lead to discussion on proper line which will be evident from positive responses to your questions.

## 3.03 Leading the Discussions

Once the discussion gets going, it is your job to stimulate and guide each member in clear thinking and understanding. You should not discourage any talkative member or directly attach a less articulate member as if he is not 'showing' any interest in the deliberations. Discussion from the talkative participant can be moved from him by asking a direct question of some other member. On the other hand, the less articulate member can be encouraged to participate in the discussion by directing the questions to such a member. In fact, there are many ways of doing this and you should use your own judgment to suit the situation.

You should keep close watch on faulty reasoning or sweeping generalisation such as "nobody is interested in development work in this company" or "you cannot get the work done from the workers these days." The way to deal with such a statement is to ask the group if that has been everybody's experience. A common sense response from the majority of the group members is most effective reply.

You should avoid the use of jargon and be prepared to clarify the meaning by re-wording the statement in simple language. Be in readiness to distinguish between the fact and opinion, and objective and subjective statements. It may be a "fact" that I started the session 10 minutes after the schedule time, but to say that I started the class "very late" is a matter of opinion. A fact is not open to argument but its interpretation can be a subject for discussion. Many a time you may be tempted to give your own opinions and views but that is very risky. It may involve you in endless argument or futile cross talk, both of which are undesirable.

## 3.04 Controlling the Trend of Discussion

If the discussion trends to narrow down to a single point of view, you should ask leading questions and pave the way for opening wide the field of discussion to accommodate other view-points also. However, you should ensure that the discussion is always kept and run on the track but not dragged. If it appears to wander, firmly but tactfully bring it back to the subject. This can be done by telling the member "you appear to have a good point there but at the moment we are discussing -------------------------," or "Mr. ---------------------- seems to make a useful contribution. May I request him to remind me about that point at appropriate time." Such statements usually have a desired effect.

Notice how the members change their opinions as the discussion proceeds. Without making direct reference to such changes, use them to help you in reaching the desired conclusions more quickly.

## 3.05 Summing up

Once a point is discussed in all aspects, you should review the points and sum up the conclusions reached. The summary should be objective and leave the group with clear idea of what has been achieved.

## 3.06 Handling Participant's Questions

a)  Deal with one question at a time.

b)  Do not allow interpretation or re-wording of a member's question by the other members

c)  Discourage interruptions and arguments, tactfully.

d)  Understand the question fully by getting the facts cleared.

e)  Pass the question to the group, making sure that they understand it as intended.

f)  Summarise the group opinion as a reply to member's question.

# 4. USING THE MANUAL

To use the manual successfully, you must understand the portion on which you are working so that the main theme can be stressed in several ways. e,g, by inflexion of voice or use of deliberate pauses, or exaggerated facial expressions and body actions. This requires good practice which comes by going through the manual several times and practicing the methods of 'stressing' in privacy. While using this manual, use pauses to glance down and get the sense of next sentence or paragraph in the manual. Do not readout word by word from the manual. In fact let your manual be inconspicuous.

The 'Procedure' column in the manual gives details of or clues to, the things to be done by you, and the 'Statement' column tells you what you should state or describe. Both these parts follow a definite sequence, and any slip or mix-up will lead to confusion, something like putting on the shoes first and then wondering what to do with the pair of socks. To avoid such embarrassments, I repeat, know your manual thoroughly and organize your thoughts sequentially.

N.B. :-  Abbreviations in 'Procedure' column:

*CB stands for chalk board.

*FC stands for Flip Charts/transparencies for overhead projection or slides.

*HO means handout

*OH Projector : means overhead projector

*mnts. Means  minutes.

*Col. = Column.

*No./no. = Number.

*QC = Quality Control.

*QA = Quality Assurance.

*Mnt. = Minute.

# 5. SESSION–WISE THINGS YOU NEED

For all sessions:

    (a) Ample supply of white and colour chalks and 2 quality dusters.

    b) Arrangement for tea/coffee/beverage in the recess during every session.

| Day/ Session | Flip Charts/ Transparencies or slides | Handouts (as part of the course material.) | Handouts (for general reading) | Other Materials |
|---|---|---|---|---|
| 1st Day Session I | FC-01 to FC-07 | VA-01 | | − About 50 copies of inaugural address<br>− Name boards for guests on dias & participants; Name strips with clips.<br>− Hints for Management introduction.<br>− Hints for comments on unnecessary cost (Cf.VA-01) Pages of manual.<br>− OH projector or slide projector, if transparencies or slides are used in place of Flip charts.<br>− DVD Player with large TV screen. (Optional) |
| 1st Day Session II | FC-08 (p.3 OF VA-02) FC-09 (p.4 OF VA-02) FC-10 (p.5 OF VA-02) FC-11 (p.6 OF VA-02) | VA-0 2 (tables on Switch Gear case) VA-03 | VM-01 VM-02 | − Sheet of paper on which "process and assembly" of 33 KV breakers is copied from Pp. 46–47 |
| 1st Day Session III | | (Visit to works/ VE Exhibition or film Show) | | − Points for lecture from pp 60–63<br>− Copy of write-up on games-Page 257.<br>− Film Projector or equivalent device |

| Day/ Session | Flip Charts/ Transparencies or slides | Handouts (as part of the course material.) | Handouts (for general reading) | Other Materials |
|---|---|---|---|---|
| 2nd Day Session IV | FC-12 FC-13 | VA-04 | | – OH projector or slides projector if transparencies or slides are used in place of FLIP charts or electronic devices with large TV screen. |
| 2nd Day Session V | FC-14 FC-15 | VA-05 | | – Copy of Specimen of information flow diagram (appended at the end of session notes as Page 95)<br>– OH projector or slides projector if transparencies or slides are used in place of Flip Charts or electronic devices with Large TV screen |
| 2nd Day Session VI | | VA-06 to VA-09 | VM-03 | – VA-08 –(A 2-pin plug & screw driver)<br>– Collect details of at least one project from Group. (Page 81. procedure column). |
| 3rd Day Session VII | | | VM-04 To VM-08 | – Prepared card containing 3 or 4 warm up exercises on creativity. Pp (124–140).<br>– Giant size: 4 wooden pieces of unequal length; Say 25 x 25 cross section. (to represent match sticks)<br>– 7 coins (p.128) Procedure column,). |
| 3rd Day Session VIII | | VA-10 (p.1,2,3 + Annex.1) | | – Members problems collected at the end of Session VI;<br>– Sufficient copies on 'information' of members problems;<br>– Your own idea lists on members problems;<br>– Pre-prepared card containing "Spur Questions" and "Strokes";<br>– Film on Creativity.<br>– Film projector. |

| | | | | |
|---|---|---|---|---|
| 4th Day<br>Session IX | | VA-12/IX | | – 2 Lists of existing ideas as given in Case Study of VA-10 as well as your own new ideas. (Paper Reel Core.).<br><br>– 2 Lists of existing ideas on participants problems;<br><br>– Pre-prepared card listing "criteria";<br><br>– Pre-prepared card containing completed matrices. |
| 4th Day<br>Session X | | VA -10<br>(remaining sheets) | | – Material for preparation of Charts<br><br>e.g.: Drg, sheets, poster colours, sketch pens, marker pens;<br><br>– Invitation cards for presentation/ Seminar;<br><br>– Prizes to be distributed. |
| 5th Day<br>Session XI | | | VM-09 | – "Guide lines for writing VA/VE report"(cf. P.5 [Para.2.13(b)])<br><br>– Pre- Prepared card ("Ask for help," etc.) pages 179–183 Procedure column;<br><br>– Schedule of seminar, if planned. |
| 6th Day<br>Session XI | | | | – Script for your Welcome Address;<br><br>– Address by Chief guest (About 50 copies);<br><br>– Script for 'Vote of thanks';<br><br>– Dinner or high tea, if planned. |

# 6. A SUGGESTED TIME SCHEDULE FOR VE/VM WORKSHOP

| Day | 09.00 to 10.30 | 10.30 to 10.45 | 10.45 to 12.15 | 12.15 to 13.45 | 13.45 to 15.00 | 15.00 to 15.15 | 15.15 to 16.30 |
|---|---|---|---|---|---|---|---|
| 1st | Inauguration by Management representative<br>Session- I<br>– Concept of Value | C<br>O<br>F<br>F<br>E<br>E<br><br>B<br>R<br>E<br>A<br>K | Session –I (contd/)<br>– Classes of Value<br>– causes of unnecessary cost | L<br>U<br>N<br>C<br>H | Session –II<br>– Job Plan Phases | C<br>O<br>F<br>F<br>E<br>E<br><br>B<br>R<br>E<br>A<br>K | Session –III<br>– Importance of team work    OR<br>– Film on over-view of VA/VE |
| 2nd | Session –IV<br>– Selection Phase | | Session –V<br>– Information phase<br>(Data collection) | | Session –VI<br>– Function/ Cost Analysis | | – Address by Guest Speaker    OR<br>– Film on Function, Cost, Worth |
| 3rd | Session –VII<br>– Speculation Phase –I Creativity | | Session –VIII<br>– Speculation (2)<br>Brainstorming<br>Members jobs | | – Visit to other industry | | – Visit to other industry<br>Film on CCeativity or C.P.S<br>(Creative Problem Solving) |
| 4th | Session –IX<br>– Evaluation Methods | | Session –X<br>– Evaluation practice thru' case studies | | Section- XI<br>– Discussion on last 3 phases of job plan | | – Visit to VE exhibition Or<br>– Film show on VE studies |
| 5th | – Address by guest Faculty | | – Syndicate work (Members jobs) | | – Syndicate work (Members jobs) | | – Brief presentation of case studies (one per group);<br>– Briefing for tomorrows session |
| 6th | Session-XII<br>Seminar- Preparations<br>Exhibiting posters & charts; trying out slides & trans-parencies; Finalisation of VE report | | Session-XII (contd)<br>– Seminar: Presentation of projects | | – Seminar (if required.<br>(Seminar may be extended in this period) | | 19.00 to 20.30<br>– Valedictory function (Dinner with spouses) |

# 2

# VALUE ENGINEERING WORKSHOP

SESSION - I

– CONCEPTS AND CLASSES OF VALUE

– DEFINITION OF VA/VE/VM

# VALUE ENGINEERING WORKSHOP

## SESSION-I – CONCEPT AND CLASSES OF VALUE
## – DIFINITION OF VA/VE/VM

| Procedure | Statement |
|---|---|
| **SESSIONS OUTLINE** | |
| 1. INAUGURAL ADDRESS BY MEMBER OF MANAGEMENT. | |
| 2. CREATING THE INFORMAL ENVIRONMENT. | Thank you, Mr. --------------- (member of the Management), Mr.----------------- (organizer of the course), and --------------- (others, if any) for the word of encouragement and support in our endeavour to succeed in the task on hand.<br><br>I am informed that the participant represents various functions. Therefore, our experiences should differ sufficiently to examine each aspect under discussion from varying points of view.<br><br>The teaching and learning experience can be a superbly enjoyable and rewarding for I and you if all of us participate willingly and fully.<br><br>Let me tell you how we will work.<br><br>Our meetings will be conducted as discussion and group participation between experienced professionals.<br><br>Whenever needed, I will provide the necessary background information and leads. When a point is thoroughly discussed and the desired conclusions reached, I will summarise the points made and conclusions reached.<br><br>Let us, therefore.  be informal. Let us share our ideas freely and examine each point under discussion without reservation. |

| Procedure | Statement |
|---|---|
| OBJECTIVE OF THE WORKSHOP<br><br>EXHIBIT FLIP CHART FC-01/I ,showing the underlined matter.<br><br><br>PAUSE, read out slowly | <br><br><br><br><br>The basic objective of this programme is to help **develop the skill in achieving the necessary functions of a product or service for minimum cost without detriment to quality, reliability and performance**, i.e. to optimise the value of an item or service. Please underline the whole sentence, and repeat with me.<br><br>Now the development of any skill requires systematised practice.<br><br>Therefore, as soon as we have agreed upon the correct approach to a particular technique, we will put it to practical test using different demonstrations.<br><br>These demonstrations will be from your own departments, and here your co-operation is 'sin-qua-non' for the success of the programme.<br><br>I shall be requesting each of you to please provide a right kind of job for our discussion and practice in the class room.<br><br>I will, of course, give you all details of such requirements as and when they arise. Your contribution will be of great value to our programme. |
| CHALK BOARD (CB)<br>"VALUE" | <br>I said "VALUE". Yes, let us first try to understand the meaning of Value. |

| Procedure | Statement |
|---|---|
| **CONCEPT OF VALUE** | |

| Procedure | Statement |
|---|---|
| Note how it is: <br>– difficult to define <br>– different things to different people <br>– confused with cost/price <br><br>DRAW OUT AND RECORD ON C.B. <br>– It is related to <br>  * Utility/need <br>  * Perform some function <br>  * Desirability <br>  * Esteem/Prestige <br>  * Returns <br>  * Economical <br>  * Satisfying wants <br>  * Scarcity <br>  * Market Value <br>  * Convenience <br>  * Good appearance/smartness <br>  *Performance (reliability) <br>  *Time <br>  *Place <br>  * Quantity <br>  * Customs <br>  * Social Norms <br>  * Sex <br><br>Use these or similar examples to bring out: <br>  *Utility/desirability <br>  *Scarcity <br>  * Esteem/prestige <br>  * Resale value <br>  * Performance <br>  * Time <br>  *Place <br>  *Quantity | What is your idea or concept of Value? <br><br>What aspects or attributes or characteristics make a thing valuable? <br><br>Air and water are very essential for supporting lives but do we treat them like valuable things? <br><br>What about a television set or a gold chain? Did our grandparents have them? Are they essential for sustaining our lives? Then why are they valuable now? <br><br>If any watch can show time – correct time – why some people go in for watches with gold plated cases and straps or digital watches? <br><br>Think of many other items of daily use and occasional use: <br>– Floral decoration on cups and saucers. <br>– Coal or gold in mines have no 'Value'. When and why they acquire different values? <br>– Painting on walls or 'panch-dhatu' idol in drawing room? <br>– A bottle of water in Sahara or a litre of petrol for a moped owner in India and vice versa? <br>– A match box for a chain smoker or for an astronaut on the moon? |

| Procedure | Statement |
|---|---|
| Additional questions are appended at the end of Session I Notes. (P.34) | |
| | Given a choice which will you prefer and why? |
| |    –  Kawasaki or Hero Honda Motor Bike |
| |    –  Motorola or Samsung mobile phone? |
| |    –  Cotton clothes v/s polyester fabrics? |
| |    –  Water or Lime juice in summer or hot cup of tea? |
| |    –  A diploma in Engineering or P.G. Degree in History? |
| PAUSE: | I may mention here that the word value is derived from the Latin word 'valere' meaning merit or worth. |
| By Lecture method using the Flip Chart FC -02/I | To summarise our discussionL Goods or services acquire value or worth when they have : |
| |    –  Capacity to satisfy want (utility) and |
| |    –  Are hard-to-get or difficult in attainment (scarce). |
| FROM FLIP CHART FC-02/I | And both the conditions must be satisfied simultaneously. |
| Economic Value: | |
|    –  Desirability | |
|    –  Scarcity | Water and air are not only desirable but essential for our survival but have no economic value because of their abundant availability. Reverse is the case with gold. |
| It varies with | |
|    –  Performance | |
|    –  Place | |
|    –  Time | |
|    –  Quantity | |

| Procedure | Statement |
|---|---|
| "Value is the least cost that can accomplish reliably a function or service"<br><br><br><br><br><br><br><br>PAUSE for the obvious answer "NO".<br><br><br><br><br><br><br><br><br><br><br><br><br><br><br><br><br><br><br>C.B. (2<sup>nd</sup> Column)<br><br><br>VALUE $= \dfrac{\text{Worth to you}}{\text{Cost you pay}}$ | We have also seen that the 'desirability' or capacity to satisfy what varies with:<br><br>– performance<br>– place<br>– time<br>– quantity<br><br>Can you measure the amount of satisfaction? Can we quantify the utility of something?<br><br>Thus the capacity of a thing to satisfy our want is **subjective** -- a matter of judgement, and it varies from person to person, from time to time.<br><br>Although satisfaction, desirability and utility are immeasurable, still we feel and say whether or not we get our money's **worth** in acquiring something.<br><br>Many a time we say or we hear people say "it is or it was worth spending a few rupees more" on such and such thing.<br><br>Presenting good value means one is pleased with what one has obtained for the money spent – in comparison with previous experience (of self or others).<br><br>Such statements are indicative of two aspects of value<br><br>One : worth, and<br>Two : cost or price<br><br>– **Worth** of purchasing or acquiring something for the price paid **(cost)**. |

| Procedure | Statement |
|---|---|
| | The attribute of satisfying want is appraised as "Worth" which may I remind you, is a subjective aspect. And the scarcity and difficulty in attainment is an **objective** aspect and expressed in monetary units as cost. |
| | The value of the industrial products, with which we are mainly concerned, is determined by this relationship of <u>worth</u> to <u>cost</u>. |
| |    – Worth which confirms to the wants of customers and the cost which matches to the resource of the customer in a given situation. |
| | A customer doesn't buy a product for the sake of buying. He buys a thing only after he is satisfied about the role or function of the product. If the purchased product does not live up to his expectation, may be the input consumption of energy is very high or operation is complicated or spares are to be specially ordered, the customer may consider that the product is of poor value. |
| | For a customer, value is measure of his satisfaction, with goods or services purchased in terms of their quality, reliability and price (consumer's angle). |
| | Producer of the goods has, in addition to the above, to consider and satisfy four groups of people: |
| |    – Customers |
| |    – Shareholders/Investors |
| |    – Board of Directors |
| |    – Employees |
| | 'Worth' being subjective, the above ratio indicates that value can't be measured in absolute terms. |
| | However, if two things give the same or almost same amount of satisfaction, or perform the same function with equal efficiency and effectiveness, we can compare their values by comparing the prices we pay for achieving the desired results. |

| Procedure | Statement |
|---|---|
| **VALUE RATIO:**<br><br>– Explain by taking a simple example (kerchiefs, door knobs, industrial fasteners, paper weights, ash-tray or the like objects). One example is given in the text. Design other examples on similar lines.<br><br>– Describe with the help of a pictorial chart : FC -03/I | |
| | Let us take an example to illustrate the point under discussion. |
| **FLIP CHART FC-03/I** | You want to buy a refill for your jotter pen. The retailer shows you two types of refills: 'Ramson' and a 'Susan' make. |
| | Ramson refills are all metallic and chrome plated, whereas the Susans use polythene tubes for barrels (ink holders). On close examination you notice that both ball holders bear the marking of a third firm with a same quality number. |
| | You are also told by the retailer that both the firms buy their ink requirements from a reputed ink manufacturer. |
| | Which one would you buy and why? |
| – Allow brief discussion.<br><br>– Furnish additional information as required. | |
| | Let me add that Ramson refill costs Rs.8/- each and Susan's Rs.5/-. each. |
| | Why do you want a refill in the first place? |
| PAUSE | |

| Procedure | Statement |
|---|---|
| | The basic purpose of a refill is to provide means of writing (with some amount of ease). Now, the writing comfort depends on the quality of the ball mechanism and ink, and in both the cases, these two parts are equally good. In other words, the capacity of satisfying wants or of performing the given function is identical in both the refills. |
| | Our choice for Susans refill is guided by the fact that for achieving a given function anything that cost less, fetches us a better value: |
| | Am I right? |
| PAUSE for the obvious reply 'Yes'. | |
| | Let us denote the worth of the function or the product worth, which is the same in both the designs, by W. By definition: |
| CHALK BOARD (C.B) | $$\text{Value} = \frac{\text{worth (of the function)}}{\text{Cost (of the function)}}$$ |
| 2$^{nd}$ column | |
| $$V = \frac{W}{C}$$ | i.e. $$V = \frac{W}{C}$$ |
| | Denoting the values and costs by $V_1, V_2$ and $C_1, C_2$ we get |
| $$V_1 = \frac{W}{C_1} \quad \text{and} \quad V_2 = \frac{W}{C_2}$$ | $$V_1 = \frac{W}{C_1} \quad \text{and}$$ $$V_2 = \frac{W}{C_2}$$ |
| $$\therefore \quad \frac{V_1}{V_2} = \frac{C_2}{C_1} = \frac{5}{8}$$ | $$\therefore \quad \frac{V_1}{V_2} = \frac{C_2}{C_1}$$ Substituting the given prices |
| $$\therefore \quad V_2 = 1.6\,V_1$$ | $$\frac{V_1}{V_2} = \frac{5}{8} \quad \text{or} \quad V_2 = \frac{8}{5}\,V_1$$ |

| Procedure | Statement |
|---|---|
| **C.B.**<br><br>Value ratio<br><br>$= \dfrac{\text{New value}}{\text{Old value}} = \dfrac{\text{Old cost}}{\text{New cost}}$<br><br>Point at the equation<br>V=W/C and bring out<br><br>– By improving worth of function (quality)<br><br>– By reducing cost<br><br>– By combination of the above. | These relations help us in comparing values of two items/products or the same product before and after VA, which achieve more or less the same function.<br><br>In general,<br><br>$$\frac{V_1}{V_2} = \frac{C_2}{C_1}$$<br><br>This ratio of old value to new value is called value ratio:<br><br>$$\text{Value ratio} = \frac{\text{New value}}{\text{Old value}} = \frac{\text{Old cost}}{\text{New cost}}$$<br><br>The above example (i.e. VR = 5/8 VS) should scatter the mistaken notion that anything which costs more is worth more.<br><br>The ratio V = W/C gives some clues for enhancing the value of a product. Can you suggest some for increasing value?<br><br>To summarise our discussion on value, value is a sort of measure of customer satisfaction, **"Value is the least cost that can accomplish reliably a function or service,"** which implies that in achieving reduced cost, performance and quality of the product should not suffer.<br><br>Now, quality and performance are not independent of each other. If performance is good, obviously quality is good. |

| Procedure | Statement |
|---|---|
| <u>EXHIBIT FLIP CHART FC -04/I</u><br><br>Explain by lecture method.<br><br><br><br><br><br><br><br><br><br><br>POINT TO FC-04/I | Let me illustrate the relation between cost and quality by means of a graphical representation. It will be seen that point 'B' represents the point of lowest cost corresponding to a certain level of quality or reliability.<br><br>If quality is reduced any further, costs will rise due to such factors as work rejection, excessive customer complaints, field service costs, consequent loss of market/less demand, etc.<br><br>For any customer there is a minimum acceptable level of quality and a top level of quality needs (points B & C on the curve). He will refuse to accept the goods below the point B and any attempt to reduce cost below 'B' will be pointless.<br><br>He will, of course, be happy if the supplier provides goods of better quality at the same price. If this is done, the supplier is the loser.<br><br>From this graph, prima-facie, it appears that our efforts should be aimed at achieving point 'B' but it is not so.<br><br>Firstly, the total cost curve is reasonably flat at the minimum cost point 'B'. Hence with a marginal increase in cost, relatively larger increase in quality will be achieved.<br><br>Secondly, the definition of value indicated that cost must be zero when the quality is worth zero.<br><br>The optimum value is obtained when the ratio<br><br>$\left\{ \dfrac{\textbf{worth/quality}}{\textbf{Cost}} \right\}$ is mazximum. |

| Procedure | Statement |
|---|---|
| | Such a point 'C' is obtained by drawing a tangent to the total cost curve drawn through the origin, 'A'. |
| | Our efforts will be to keep the total cost towards the point 'C'. If the actual cost is to the right of 'C', efforts will be made to bring it down and if it is towards left of 'C', we will aim at bringing it up to 'C' |
| | Consequently, depending upon the individual cases, the total cost may either go up or down in achieving the optimum value. |
| | Thus, the techniques we are going to learn and apply to live cases are not necessarily a part of cost reduction schemes. Our aim is to optimize value. |
| | Then, how to go about determining the right composition of quality, reliability, efficiency, price and service which will ensure maximum satisfaction to purchaser at the ultimate economy to the supplier. |
| | Techniques of determining this ideal combination is called value Analysis or value Engineering. We will come to this subject after a while, in second Session. |

| Procedure | Statement |
|---|---|
| **CLASSES OR TYPES OF VALUE** ||

| Procedure | Statement |
|---|---|
| By lecture method:<br><br>Go to CB and score out words as they occur in narration:<br><br>  – Want<br>  – Desirability<br>  – Scarcity<br>  – Performance<br>  – Time<br>  – Place<br>  – Quantity, etc.<br><br>and relate the remaining words to<br><br>  – Use value, followed by<br>  – Esteem value,<br>  – Exchange value,<br>  – Cost value.<br><br>Chalk Board:<br><br>  – Use Value<br>  – Esteem value,<br>  – Exchange value,<br>  – Cost value. | Let us go back to your contribution to the concept of value and see what has been summarised on the board work and charts and what is left out.<br><br>First, there should be want or desire to possess a thing which depends on individuals' needs, taste, religious beliefs, customers, sex etc.<br><br>Secondly, the thing should be difficult to get or attain it.<br><br>We have also seen that the degree of desirability coupled with scarcity of a product or service varies with its performance (water & lemon juice), time (summer or winter; day or night; torch or headlight);<br><br>Place (Sahara or Siberia; water & fridge); quantity (teaspoon full or tub full of medicine/water for bathing).<br><br>Lastly, goods acquire value partly because of their ability to perform use functions reliably compared to the price paid and partly because of esteem or prestige attached to their possession- concern that they should look smart and pleasing.<br><br>Again, there is the resale or exchange value which allows trading of the item for something else.<br><br>Finally, the sum of labour, material and overhead costs required to produce something becomes its cost value.<br><br>The sum of these four values is called the total or economic value. It may be emphasized here that percentages of these components in total value changes from item to item, time to time, etc. |

| Procedure | Statement |
|---|---|
| CLOSE FLIP CHART FC -05/I | First two values are concerned with physical properties and the last two are related to economic characteristics.<br><br>The different classes of value are defined as follows:-<br><br>1. Use Value :  Properties that accomplish a use, function, or service.<br><br>2. Esteem Value :  Properties, features or attractiveness, aesthetic appeal, etc., which make the ownership desirable and motivates people to buy it in preference to other products or equipment.<br><br>3. Exchange value :  Properties or qualities which enable the object to be exchanged for something else.<br><br>4. Cost value :  Sum of labour, material, and overhead costs required to produce something.<br><br>For value Analysts/Value Engineers, use and esteem values are of particular significance, use value being of paramount importance.<br><br>In our deliberations all other values will be subordinated do it in varying degrees. |
| Put off OH projector.<br>Flip charts. | |

| Procedure | Statement |
|---|---|
| **CAUSES OF UNNECESSARY COSTS** ||
| HANDOUT: VA-01/I (P.215)<br><br>Distribute sheets.<br>Say as you go round. | <br><br>Since we are concerned with elimination of any cost which does not add to value, for the next few minutes, we will consider how the unnecessary costs creep in.<br><br>A number of causes have been identified and classified by the experts in the field.<br><br>To help you to remember them, they are listed on this sheet. |
| Sit down, read items, amplify those appropriate to the group. (Your sheets cf. PP 32–33 have the necessary hints for amplification. Invite illustrative examples against each factor). | <br>I hope it is quite comprehensive list and in your varied experience you may have detected such problems of unnecessary costs. |
| – Allow time for consideration of all the causes listed on the sheet.<br>– Get few examples & relate them to appropriate headings. | Let us spend a few minutes looking at these in order to see whether we have come across them anywhere at any time.<br><br>Does this discussion show ways of avoiding unnecessary costs? How? What should be our approach to value problems? |
| PAUSE | |

| Procedure | Statement |
|---|---|
| Don't record anything on C.B. Don't comment on the contribution at this stage. Allow about 5 to 10 minutes. | All I can say at this stage is that there should be some logical and systematic method for analysis and for solving the problem of unnecessary costs, i.e. to achieve optimum value.<br><br>As we have seen earlier, in doing so, sometimes the cost may go up marginally for achieving a relatively larger increase in the worth.<br><br>Our emphasis is on elimination of unnecessary cost, and not on mere cost reduction. |
| **Value Engineering/Value Management** | |
| | Like any other systematic method VA/VE/VM depends on having a plan of dealing with projects and in VA/VE, this systematic and logical method is called **JOB PLAN**<br><br>Before taking up the plan, let us first consider the terms:<br><br>    Value Analysis/VA,<br><br>    Value Engineering/VE,<br><br>    Value Administration,<br><br>    Value Assurance, etc.<br><br>Please recall the basic objective of our course, which you repeated with me and which was to '**achieving the necessary function of a product or service for minimum cost without detriment to quality, reliability and performance**,' i.e. to optimize the value of an item or service.<br><br>A disciplined or organised procedure of developing this skill is VALUE ANALYSIS. |

| Procedure | Statement |
|---|---|
| EXHIBIT FLIP CHART FC-06/I | Since this organised procedure is applicable to any product or service, it acquired different titles depending on the functional areas where this technique is used.<br><br>The application of the ideas of value Analysis to the **prevention of unnecessary cost is called Value Engineering.** It (VE) is the application of VA techniques in the **design and development phases.**<br><br>V.A. is a **remedial** process and VE is a **preventive** method. Both are disciplines using the same techniques and differ only in their phases of applications.<br><br>Although initially the techniques were used mostly in the Engineering fields which gave it its name (VE), it can be, and is now, used in many other areas like marketing, office work, purchasing, hospitals, etc., and a more general term, V.A. is now used for it.<br><br>Value Administration is the group of the same Techniques applied to administrative functions and so on.<br><br>The latest term is "VALUE MANAGEMENT" (VM) and covers all "COST" areas.<br><br>In our discussions, VA, VE, and VM will be used synonymously.<br><br>Call by any name, it is a multipurpose tool like this. |
| EXHIBIT FLIP CHART FC-07/I | Before concluding the session let me emphasis that in VE/VM exercise, we don't question the function or try to eliminate it.<br><br>Our efforts are directed towards achieving that function at the least cost without detracting from quality, be it a product performance or service function. |

| Procedure | Statement |
|---|---|
| | Therefore, the example that you may provide for our exercises need not confine to Design or Engineering aspects.<br><br>They can be from various functional areas like Purchase, QC, Utility Services. Administration and so on, where 'costs' are incurred. |
| Close session with appropriate remarks like "Thank you for your contribution to the discussion."<br><br>Announce schedule of 2$^{nd}$ session.<br><br>Close charts.<br><br>Clean C.B. | |

| Participants' copy contains | Hints for your comments |
|---|---|
| **STIMULATING DISCUSSION ON** <br> **CAUSES OF UNNECESSARY COSTS** ||
| (Participants' copy of hand out VA-01/I contains the following) | <u>Hints for your comments:-</u> |
| 1. <u>MANAGEMENT INEFFICENCY:</u> | "That will do" (Chalta Hai) attitude. Belief that normal cost reduction activities will reveal all unnecessary cost, then why VA/VE? |
| * Complacency: (Routine cost reduction schemes will take care of unnecessary cost. Then why VE?) | This is not so; VA/VE is more searching. |
| • <u>Lack is value objectives</u> <br> – Goals Vs Value <br> – Business pressures | No thought given to goals relative to value – goals like enhancing value by 5% or 10% p.a., belief that business pressures force changes without reference to costs. Then why bother to change through VA/VE. <br><br> The foregoing faults manifest in this form. **<u>If you fail to plan, you are planning to fail.</u>** |
| * Lack of Planning <br> * Lack of Pressure <br> * Lack of knowledge | To achieve value objectives in the form of target cost add to unnecessary cost. <br><br> As to how the value problems arise and to how to deal with them. |
| 2. <u>INABILITY TO APPLY VA/VE:</u> | <u>This engenders unnecessary costs in many ways:</u> |
| • Lack of (technical and cost) information. <br><br> (– Full and accurate information helps in taking meaningful decisions. <br><br> – Information has to be continuously updated.) | Designer may not be aware of availability of a new material, process or machine which enables him to produce better design. Similarly, Production Engineer may not realise that a design change can be made to enable a new and better process to be utilised or Purchase Office may be unaware of superior substitutes for materials/vendors/suppliers. |

| Participants' copy contains | Hints for your comments |
|---|---|
| * Lack of communications— | Poor communication results in waste of time efforts which cost money resulting in unnecessary costs. Root cause of this is laziness and timidity. |
| * Lack of good ideas | |
| – Available talent is seldom fully utilized. | |
| – Many good ideas are still-born because they never recorded or developed. | Chrome- plating is peeling off- so increase thickness of coating (high cost) or reduce dia. of the pipe (strength reduces). But why not think of using polyvinyl ABS material? Esteem value function of chrome plating can be satisfies by P.V. material, strength and anticorrosion properties are not affected. |
| * Lack of training— | ...in understanding how the value problems arise and how to deal with them. (This provides the discipline and opportunity for evolution of new ideas, their evaluation). |
| 3. HUMAN WEAKNESSES: | All humans are prone to weaknesses. |
| * Honest wrong beliefs.— | Harmless ones like the world is flat or God has 3 heads and 4 hands or no form at all. |
| | Harmful and costly ones like Titanium can't be cast, now being done effectively and economically. |
| | – That glass is too brittle to use as a structural material (R.G.F. is now a commercial material). |
| | – That "Atom can't be split" is another such example. |
| | – That stainless steel is non-magnetic. |
| * Habits and attitudes:- <br> - Reluctance to accept new ideas. | ... acquired during their lives. Most people resist change and technological innovations call for change to be competitive. |
| - Obstructive and negative attitudes. | And new ideas do carry with them the smell of unknown danger. |

| Participants' copy contains | Hints for your comments |
|---|---|
| \* Playing safe: <br><br>   – Duplicating safety features. <br><br>   – Finer tolerances than required. <br><br>   – Excessive packaging <br><br>   – Over-design, etc. <br><br> \* Pride in own ideas........ <br><br><br> 4. <u>Competitive pressures</u> <br><br>   – Anxiety to beat competition. <br><br>   – Lack of time. <br><br>   – Over confidence. | – Exaggeration of environmental factors and specifying a strong or more material than required <br><br> – specifying better finishes. <br><br><br> .....Deliberately looking for faults in the proposals to condemn them is the result of excessive pride in their own ideas. <br><br><br><br> Promising to deliver a new design by a particular date when design is still in embryo stage do beat competition; result-complicated design – high costs. |

## Additional Questions

– Refrigerator to an Eskimo or to an urbanite in Delhi?

– Imported jean to a mod girl or her mom?

– Regular supply of LPG cylinders to an urbane eve and of fire wood to a village woman and visa versa?

– Why a big dog-ear like collar to a short? What was its origin?

– Gold Zari (laced) turban to a modern groom compared to a prince of the by-gone era?

– Parker pen to an illiterate farmer and sickle to a Pandit?

– What about old postage stamps and coins?

– Common stone, black granite and marble?

– Car fender ..chrome plated. Why ?

– Woolen suit for a Sanyasi/top flight Western Executive?

– Sugar candy for a child/diabetic patient?

– Stainless steel v/s brass utensils?

– A capsule of medicine: a quintal of rice and vis versa?

– Tumbler full of water or drum full water when you are thirsty?

– Plastic v/s porcelain crockery?

– Lipstick or saree for a young man?

– Flying a bomber plane or driving a moped?

– Why ornamental handles for doors or figure heads (Centaur or horse) on handles of bonnet?

– Why multi colour printed tins sheets used for 'cans' of chocolates?

# 3

# VALUE ENGINEERING
# WORK SHOP

## SESSION - II
## — JOB PLAN PHASES

# VALUE ENGINEERING WORKSHOP

## SESSION - II: - JOB PLAN PHASES

| Procedure | Statement |
|---|---|
| OPEN SESSION:<br><br>Make suitable opening remarks: Groups participation etc. Stress that success of the value Workshop depends on their contributions. | |
| **REVIEW SESSION – I** | |
| If the session – II is conducted on the same day, skip off this REVIEW | Let us briefly review our last session. You will remember that something acquires worth when it has some utility. Utility is a necessary condition for value, but not sufficient one (air, water).<br><br>There should be also scarcity. We have seen how the concept of worth or quality differs from person to person and how value with:<br><br>- Performance<br>- Place<br>- Time<br>- Quality<br><br>A little later we considered four types of value:<br><br>- Use or function value;<br>- Esteem value;<br>- Exchange value; and<br>- Cost value.<br><br>We also recognised that value is the least cost that can accomplish reliably a function or service. |

| Procedure | Statement |
|---|---|
| | From your varied experience you were quick enough to endorse causes of unnecessary costs listed on a sheet (VA-01/I) |
| | After looking at these causes, our discussion turned to achieving optimum value through the Techniques of VE/VM. |
| | VE/VM was defined as an organised method of identifying and eliminating unnecessary cost, without diluting or compromising on quality, reliability and delivery schedules. |
| | In VE/VM exercises we don't question or eliminate function but direct our efforts towards achieving that function reliably by consuming least resources. |
| | You will remember that VE/VM is a 'cost avoidance' techniques unlike other cost reduction techniques which aim at "correcting the cost". |
| | VE is a more fundamental approach which takes nothing for granted and attacks everything about a system including the existence of the system itself. |
| | Let us now go a step further into how to attack the problem of optimizing the value of a product or service. |

| Procedure | Statement |
|---|---|
| **Recognition of Steps in Job Plan** ||
| PAUSE: | Let us now see how a group of Value Analysts working in a multi-product organisation looked at their job and conducted a value analysis session. |
|  | The team consisted of the five persons: Messrs Gupta, Raman, Pai, Sufi, and Bright. |
|  | Could you guess which departments they represent? |
|  | Well, |
|  | Mr. Gupta is from marketing; |
|  | Mr. Raman from Accounts/costing? |
|  | Mr. Pai from Switch Gears Design; |
|  | Mr. Sufi from Purchase; and |
|  | Mr. Bright from Methods Dept. |
|  | Mr. Bright, a value specialist, is the group leader for the project. Let me inform you that the company manufactures power distribution equipment like transformers, circuit breakers and control panels. |
| Allow few moments to absorb "Who is Who". Check, understanding. Let is sound "Casual" and not as test of their memory power |  |
|  | I would like you to please concentrate on the team's approach to the value problem, not on the style of dialogue OR technical details at this stage. |
| SIT DOWN TO READ ALOUD THE DIALOGUES | What transpired at their first meeting was something like this; |
| Let the dialogues sound natural. Vary tone, if possible. |  |
|  | Mr. Bright: "So Where do we begin our VE exercise?" |

| Procedure | Statement | |
|---|---|---|
| He is facing difficulties in procurement of air chambers in small quantities- mostly as replacement spares. | Mr. Sufi: | "Let us take some assembly of our first product Air Blast Circuit Breaker (ABCBs). Say Air Chambers. |
| | Mr. Gupta: | "Nobody is asking for ABCBs now- a days." |
| | Mr. Pai: | "let us then consider our latest product – 240 KV Oil Circuit Breaker." |
| | Mr, Raman: | "That is our bread and butter – there is no competition to us for this product. Also, let me tell you that 240 KV breakers constitute only about 12% of our Sales volume but account for about 35% of the total profit." |
| | Mr. Pai: | "Which product is then selling at least profit?" |
| | Mr. Gupta : | "33 KV Minimum Oil Circuit Breakers." (MOCB) |
| | Mr. Raman: | "Yah! I know it but why are you selling it at a loss and that too in large numbers?" |
| | Mr. Gupta: | "Well, that is a business strategy. As long as "buy under one roof" instinct or habit persists, we will have to do it. You know our customers for higher capacity breakers are asking for 33 KVs also.<br><br>As you may be aware, competitors have entered the market with a newer type of breakers serving the same purpose.<br><br>Discontinuance of 33 KV may draw away our customers for higher capacity breakers also." |

| Procedure | Statement | |
|---|---|---|
| | 2 or 3 voices | |
| | at a time: | "let us then take up 33 KV MOCB for further analysis." |
| | Mr. Pai: | "Is it the profit margin alone that qualifies a product for V.A.? I learn that the field service complaints are a bit high in case of power transformers." |
| | Mr. Bright: | "I agree. But profit margin is one of the most important criteria to start a VA program. |
| | Mr. Gupta: | "Whatever be the criteria and whichever be the product, where do we make a start? There are hundreds of parts in each product and it is impossible to tackle them in one shot." |
| | Mr. Bright: | "That is true. Why not then take up some major assembly or sub- group?" |
| | Mr. Sufi: | "On what basis? But that also is not an easy task. We must know what is costing more - say in materials or in manufacturing." |
| | Mr. Raman: | "I can furnish the material and manufacturing cost data." |
| | Mr. Bright: | "Thank you Mr. Raman. We will need your help." |
| Note: What is most tangible is offered spontaneously. | | |

| Procedure | Statement |
|---|---|
| | Mr. Pai:      "Cost alone should not be the guiding factor. Manufacturing problems or customer's requirement may not allow us to touch some parts or sub-assemblies." |
| Mark the feeling of "some one running away with credit" of suggesting the basis of VA Study. Also, fear of exposure of weakness of the Design Dept. | |
| | Mr. Sufi:      "Can you tell us what those constraints are?" |
| Note clash of professional pride. | |
| | Mr. Pai:      "How can I list them here and now? Are we going to study all the drawings and process sheets?" |
| | Having sensed that the discussion is likely to wander away, the group leader tactfully intervened, saying: "There is a point in what Mr. Pai said. It is a Herculean task but as a group, it should not be an insurmountable one. |
| Mark the moderating tone in bringing back the discussion to the points. | As suggested by Mr. Sufi, I feel we may begin by having a closer look at the material cost structure of 33 KV Breaker, which Mr. Raman has agreed to furnish. |
| | I hope, very soon we will have the opportunity to take up jobs on the basis of other criteria like field service complaints, etc." |
| | What was the problem facing the group? |
| | What was the group doing at this stage? |
| Bring out <br> "Selecting a job" | |

| Procedure | Statement |
|---|---|
| <u>Chalk Board:</u> 1ˢᵗ Column<br><br>1. "SELECTION"<br><br>Long Pause | On the following day the group assembled at the pre-appointed time and the venue. Mr. Raman provided the cost break-up of different types of materials going into the making of 33 KV breakers.<br><br>The cost structure revealed that the castings, which form 6% of the total parts by nos. account for about 33% of the total material cost. |
| <u>Distribute Handout VA-02/II</u> | This information is given on the 1ˢᵗ page of this handout. |
| Refer to Table-1 | The arrangement of cost data in descending order further revealed that the "brass cylinder" tops the lists, followed by aluminium castings for "fixing Flanges" to porcelain insulators. |
| Refer to Table – 2 | Please refer to 2ⁿᵈ page of the handout VA-02.<br><br>Mr. Bright informed that earlier a joint study was carried out by Design and Methods people on the brass cylinder and found that the present material offers the optimum value.<br><br>The group members were aware that a small percentage of saving on high cost item would yield more saving than a high percentage of saving on a less-costly item.<br><br>Hence, the FIXING FLANCE was the choice of the team for further (value) study. |

| Procedure | Statement |
|---|---|
| | Having selected the job, a spate of questions followed: |
| | "How big is the flange?" |
| | "How it is helping to join two parts together?" |
| | "What is its main purpose?" |
| | "Any machining operation?" |
| | "Why a separate flange?" |
| | "Why aluminium?" |
| | "Will the customers accept any other materials?" |
| | "Will any other material withstand the forces acting on the flange?" |
| | "What about appearance and quality? |
| | and so on. |
| | But none could answer any of these questions in definite term. |
| <u>Bring out</u><br><br>"Detailed data or information."<br><br><u>Chalk Board</u><br><br>2. INFORMATION | Why? What was the group seeking through these questions at this stage? |
| | What sort of information should they have to proceed further? |
| <u>Bring out</u><br><br>(If required, repeat the question slowly)<br><br><br>- Design specifications<br>- Functions<br>- Manufacturing. details<br>- Cost, etc. | |
| | At this stage, Mr, Bright announces the assignment schedule i.e. the collection of detailed information regarding design specifications, dimensioned drawings, operation sheet, assembly particulars, |

| Procedure | Statement |
|---|---|
|  | Cost and functional details, service conditions, suppliers, etc. |
|  | They also discussed and identified the sources from which the most reliable information could be had. |
| CHECK UNDERSTANDING.<br><br>Elaborate, if need be.<br><br>LONG PAUSE |  |
|  | The third meeting was the beginning of serious business sessions. It started with Mr. Pai's explanation of the product, its function, field service requirements, etc. with the help of drawing of this type. |
| Show drawing of 33 KV<br>Breaker: FC-08<br>(P.3 of handout VA-02) |  |
| Stand up and explain without using manual. Describe briefly the main assys., & followed by construction & function of flanges, gasket, spiral, spring, etc. | Mr. Pai :  33 KV Breaker is a triple-pole Minimum Oil Circuit Breaker for out-door erection. It comprises 3 major assemblies, viz: |
|  | No. One :  3 poles (A) fitted on to a base frame mechanism (B) with a common operating shaft & link up system enclosed therein (not shown in Fig.) |
|  | No. Two :  A spring – closing type operating mechanism housed in a cabinet (C); and |

| Procedure | Statement |
|---|---|
| | No. Three: Support structure (D) |
| | Each pole has two porcelain insulators. The upper insulator houses interrupting chamber and the lower one is a support insulator. Both are filled with oil. |
| | The arc extinguishing chamber contains a fixed contact and a moving plug contact. The operating movement from the mechanism housed in (C) is transmitted to the moving contact by a series of links, cams and gears. Schematically this is shown here. |
| EXHIBIT –FC-09/II<br>(P. 4 of Handout VA-02/II) | |
| | (UT) is the upper terminal connected to the fixed contact (FC). |
| | The lower terminal (LT) is connected to the moving plug contact (PC) via the contact guide and roller contractors (CG) |
| | PC is pulled down or pushed up by means of a FRP pull-rod (PR) passing through the hollow support insulator. |
| | This pull rod, in turm, is connected to the operating mechanism in Cabinet (C) through an operating shaft (S), an arm (A) and an another pull rod (P). |
| EXHIBIT –FC-10/II<br>(P. 5 of handout VA-02/II) | |
| | The process of manufacturing and assembling the flange is like this: |
| Copy the "Process" & "Assembly" on a separate sheet and read step by step. | |
| OBJECTIVE: To stress on fine tolerances, use of jigs & fixtures and a large number of operations involved in machining & assy. | |

| Procedure | Statement |
|---|---|
| EXHIBIT –FC-11/II<br><br>(P. 6 of VA-02/II) | PROCESS:<br><br>Part No., Drawing No., material specification, dimension pattern (rough casting), time per operation, etc. are given on the route card. The operations are as follows:<br><br>1. Hold on 212 Ø. Face flange to 20 mm thick.<br><br>2. Bore Ø 194.2 (+ 2.2 – 0.0) through hole.<br><br>3. Bore Ø 196 x 36.5 deep.  Deburr.<br><br>4. Reverse and hold on turning fixture.<br><br>Face to 56 mm, bore Ø 194.2 (+0.02).<br><br>Form internal groove Ø204.5(+0.2/-0) with R4. Groove width 9.5 (+0.3/-0.0).  Deburr.<br><br>5. Using drill jig,  drill Ø 10 hole.  Deburr.<br><br>6. Using drill jig, drill 4 holes of Ø 18.  Deburr.<br><br>7. Spot face Ø 30. Deburr.<br><br>8. Clean and degrease.<br><br>9. Paint un-machined  surface.<br><br><br><br>Assembly:<br><br><br>1. Clean insulator, keep in position and match flange-1 with it.<br><br>2. Locate gasket (G1) on base plate.<br><br>3. Position Flange-1 on base plate<br><br>4. Locate insulator in flange-1.<br><br>5. Insert spiral spring (S-1).<br><br>6. Assemble flange-1 to base plate with 4 bolts, spring washers, and nuts.<br><br>7. Match contact house with flanges-2<br><br>8. Assemble flange-2 to insulator.<br><br>9. Insert spiral.<br><br>10. Apply red oil proof enamel to Gaskets-2. |

| Procedure | Statement |
|---|---|
| | 11.  Place one of the gasket -2 on insulator. |
| | 12.  Assemble contact house. |
| | 13.  Assemble 2nd gasket-2 |
| | 14.  Locate 2nd flange-2 with 4 bolts, spring washers and nuts. |
| | 15.  Locate insulator. |
| | 16.  Insert spiral. |
| | 17.  Tighten bolts and nuts. |
| Invite questions (if any). Clarify as necessary. No discussion. | |
| | The material specifications, shape and special features of the flanges, and the spiral spring were critically examined from the functional point of view. |
| | The functions achieved by each element and cost of achieving them were found to be like this: |
| Refer to page 7 of handout VA-02/II (Cost- function Matrix) | |
| | The information was thus fully analysed and organised in a systematic manner. |
| | At this point of time, Mr. bright announced Tea Break. |
| PAUSE | |
| | Mr. bright commenced the 4th sitting by explaining as to how they would work during the next few minutes. |
| | He said: "We are about to enter a very crucial and creative phase of our deliberations, and the success or failure of our study hinges on this session. |
| | Now that full information is available at our disposal, many a question may be agitating your mind, or you are anxious/ eager to suggest better alternatives. |

| Procedure | Statement |
|---|---|
| | Before doing so, let me emphasise on a few points:- |
| <u>C.B.</u> | No.1:   We will consider only one point at a time. |
| 1. One point at a time.<br>2. The more the ideas the better.<br>3. No pre-judgement of ideas.<br>4. Note all ideas. | No.2:   All ideas are welcome – the more the better. |
| (Emphasise on ALL IDEAS – even those seemingly appear to be impractical or fantasies) | No.3:   No one should praise or ridicule or comment on, anybody else's idea – neither any evaluation of merits or de-merits of any suggestion at this stage. |
| | No.4:   I will note down <u>all the ideas</u>. |
| | The group then took up each part, each feature related to fixing of two flanges together and subjected it to close scrutiny:- |
| | -   What is it?<br>-   What does it do?<br>-   How does it do that?<br>-   What else can do that and so on. |
| | What was the group doing at this stage?<br>Why questioning? |
| <u>Bring out:</u> | |
| Finding alternatives or solutions. | |
| If the answer is "critically examining or challenging," put a counter question – "What for? What is the purpose of such examination?" This should lead to desired answer. | |
| <u>C.B.</u><br>"SPECULATION" | |

| Procedure | Statement |
|---|---|
| | They are speculating on different ways of achieving the functions listed earlier. We call this phase as "**Speculation**." |
| | The ideas produced during this sitting were listed like this: |
| Refer to page 8 of Handout VA-02/II | |
| Read out first 4 ideas | |
| | The next step was to look back at the ideas Mr. Bright has jotted down. |
| | They discussed the advantages and disadvantages of the ideas and their practicability, making sure that all good points were recognised and recorded. The agreed decisions were tabulated thus: |
| Refer to page 9 of handout VA-02/II | Why such comparative statement? What was the task before the group? |
| Bring out: | |
| Selection process of ideas/comparison of ideas. | |
| C.B. | |
| "EVALUATION" | Some ideas were dropped like hot potatoes; others were picked up for further examination. |
| | The idea which topped the list in preliminary evaluation was then subjected to close scrutiny. |
| | Every aspect of it was challenged by using check – list of the following criteria:- <br> - Design <br> - Manufacturing <br> - Performance <br> - Quality |

| Procedure | Statement |
|---|---|
| | - Marketing<br><br>- Field service, etc.<br><br>This critical examination threw up a no. of questions to be answered by experts and professionals of various disciplines.<br><br>- Who will certify soundness of the new design proposal?<br>- What data will be required before meeting the specialist?<br>- When and who will collect?<br>- Any meeting is called for? If so, who should attend? when? and where?<br>- Is it worth making a prototype?<br>  Any proving is involves?<br>- What will that cost?<br>- How long will it take?, and so on.<br>- What was group engages in at this stage? |
| Bring out:<br>Chalking out plan of action or preparing a plan for developing ideas.<br><br>C.B.<br>PROGRAM PLANNING<br><br>Bring out:<br>Take action<br>Execute the plan<br><br>C. B.<br>PROGRAM EXECUTION | Having prepared a thoroughly sound and workable plan of action, what should be the next stage? |

(Watch out for generalistic answers like 'implementation' or 'follow-up', Since the idea is not yet 'developed', question of implementation does not arise. 'Follow-up' follows 'implementation' and hence far, far away from the planning stage).

| Procedure | Statement |
|---|---|
| Sit down and read | Before calling it day, Mr. Bright invited the Value Specialist to give specific guide lines for this phase. |
| | Some excerpts from value Specialist's address to the group: |
| | "Main purpose of this phase is to develop selected ideas into practical proposals." |
| | "You are likely to encounter several difficulties and road-blocks while discussing the ideas with the specialists." |
| | "They may give many excuses to prevent changes." |
| | - Why change it when it is working satisfactorily?<br>- It costs too much to change.<br>- Customer won't accept it.<br>- Mr. G.M. doesn't like it, so on and so forth. |
| | "You must not over –react to any criticism of their ideas. Please remember " resistance to change is normal phenomenon." |
| | "Ask for help in developing ideas." |
| | "Give credit and guarantee acknowledgement for their contribution." |
| | "Give all necessary support data and allow time for their consideration." |
| | "Adopt positive attitude in questioning and also in responding to suggestions for improvement." |
| PAUSE | "Finally compile a table indicating how the developed ideas compare with the existing design,  on point – to- point basis." |

| Procedure | Statement |
|---|---|
| | "Also give estimates for time and cost for implementation, and also recurring and non- recurring saving. |
| PAUSE | |
| Change tone: | |
| | Is that clear? (Role of Value Specialist). |
| | Let us now see what happened in the fifth session. |
| | The numbers first exchanged notes on their experiences- good and bad. It was followed by the disclosure of information collected by them, the gist of which is like this: |
| | Mr. Sufi: |
| | Rs. |
| | The cost of machined GI casting   22.70 x 10 |
| | Cost of cementing flange to insulator   10.00 x 10 |
| | Apportioned cost of fixture for Cementing   <u>1.40 x 10</u> |
| | <u>34.10 x 10</u> |
| C.B. | |
| Rs. | |
| Cost of GI casting   22.70 x 10 | |
| Cementing   10.00 x 10 | |
| Fixture   <u>1.40 x 10</u> | |
| Total   34.10 x 10 | |
| Say   34.00 x 10 | |
| Extra Transport   <u>1.00 x 10</u> | |
| Total   Rs.   <u>35.00 x 10</u> | |
| | The suppliers have agreed to supply insulators with machined G.I. flanges cemented to their ends @ Rs. 340.00/ piece |

| Procedure | Statement |
|---|---|
| | However, there will be a marginal increase of Re. 1.00/flange in freight charges due to the increased weight of G.I. casting. |
| | **Mr. Pai:** |
| | "Fixing flanges of higher capacity poles are made of G.I. castings cemented to the insulators, and our proposal is acceptable to Design/Engineering. |
| | It is learnt that this change was contemplated by them also." |
| | **Mr. Bright:** |
| | "We placed an educational order on M/s Delta Castings and assembled one breaker using G.I casting flanges. The test reports are very encouraging. |
| | However, regular suppliers should use a fixture to ensure 'squareness' of the flange face and insulator axis. |
| | They have agreed to do so provided we bear the cost of the fixture." |
| | **Mr. Gupta:** |
| | "No difficulty is seen in marketing the product with G.I. castings. In fact, creep resistance of G. I. being higher than that of aluminium, it may become an additional sales feature." |
| | **Mr. Raman:** |
| | "Here is the table of cost – function comparison: |
| Refer to Page 10 of handout VA-02/II (Cost –function Matrix after VA) | |
| | Having satisfied that the outcome of their efforts was gratifying, they prepared a draft report high-lighting the advantages and envisaged changes. |
| | Later on they conducted a presentation of their findings to the top management representatives with a view to gaining overall acceptance of the developed ideas and approval for implementation. |

| Procedure | Statement |
|---|---|
| | When the change was fully implemented and use of G.I castings became a standard practice, final report was prepared, serially numbered and distributed amongst the concerned agencies.<br><br>This, in our plan of action or Job Plan, as it is called, is the last phase viz: submission of the proposal and implementation. |
| C.B.<br><br>"IMPLEMENTATION" | Well, this, in a nutshell, is then the VA/VE programme:<br><br>Many people when they are first introduced to VA/VM come up with remarks like, "we are already doing that but we don't call it value Analysis or value Engineering".<br><br>Or "This is just a common sense, why the fancy title?"<br><br>Well, to a certain extent they are right. There is very little in V.A./V.M. that is original or even very complicated in theory.<br><br>It is this very simplicity that makes it difficult to understand why the world's industrial leaders have become vitally interested in it;<br><br>Why the Government of advanced countries provide for it in major contracts;<br><br>Or why it is being taught world wide in colleges and business associations.<br><br>The answer is that it is a planned and formalized procedure, and thereby contributes over and above what the normal application of common sense has been able to accomplish without it. |

| Procedure | Statement |
|---|---|
| **JOB PLAN SHEETS** | |
| <u>Distribute</u><br>JOB PLAN SHEET<br>VA-03/II | |
| Talk while going round the table | The different phases of the Job Plan that we have identified are listed on this sheet. |
| | I would like each one of you to have this sheet. Please bring it with you to each of the following sessions. It will form the basis of our discussions and practice. |
| Sit down from here & read out from manual | The purpose of the plan is set-out at the top of the sheet: |
| | - **A practical plan for efficient identification and elimination of unnecessary costs without detriment to quality and reliability.** |
| | There are 7 basic steps to value Analysis/Value Engineering methodology. |
| | These are not always distinct and separate; in practice they often merge or overlap. |
| | For example, program planning and program execution can be merged into one phase, or Information phase may be split into two parts: viz., data collection and cost-function analysis. |
| | As practical people you will want to know how this plan can be applied to your jobs. Well, that is what we are aiming at in this VE workshop. |
| | So in our remaining sessions, we will put it to the test by taking jobs from your areas of working. For this purpose we will work in groups. |

| Procedure | Statement |
|---|---|
| <u>Announce Groups:</u><br><br>Each group to constitute of 4 or 5 members, including a "potential leader" as you might have spotted during the discussions held so far. | <br><br>Will you please therefore select one job, preferably a short one suitable for class work but which you really desire to improve in its value.<br><br>I will see each group before your job is taken up for value Analysis, so that you will be clear as to what is required and I will be knowing what the job is!<br><br>Any questions? |
| - Clarify points, if any.<br><br>- (Distribute handouts VM-01/& VM-02/II.)<br><br>- Make suitable closing remarks.<br><br>- Clean C.B.<br><br>- Meet at least 2 groups between the sessions.<br><br>- Help in choice of the job.<br><br><br><u>FILM SHOW:</u> | |

# 4

# VALUE ENGINEERING WORKSHOP

## SESSION - III
## — TEAM WORK

# VALUE  ENGINEERING  WORKSHOP

## Session - III :- TEAM WORK

| Procedure | Statement |
|---|---|
| Open the session with suitable remarks.<br><br>Chalk Board (C. B.)<br>(1st Column)<br><br>JOB PLAN PHASES:<br><br>- Selection<br><br>- Information<br><br>- Speculation<br><br>- Evaluation<br><br>- Program Planning & Execution<br><br>- Implementation<br>- Status Summary & Conclusion. | |
| **REVIEW SESSIONS I & II** | |
| | In session – I we discussed the concept of Value and decided that Value is the least cost that accomplishes reliably a function or service and that it is a relative term best expressed as a ratio of product worth to product cost or service worth to service cost.<br><br>A little later we considered the causes of unnecessary cost.<br><br>We also defined VA/VE and distinguished between VE and other cost reduction techniques. We will remember that VA/VE is a cost avoidance technique – an orderly procedure for optimizing the value. |

| Procedure | Statement |
|---|---|
| READ the job Plan phases from C.B. | In session – II with the help of a case study, we recognised the several stages of this orderly procedure called "Job Plan," which are; |
| | (Selection, Information, to Conclusion). |
| Stress on "TEAMS". | Lastly, I mentioned that like the team led by Mr. Bright, we will also work in teams or groups. |

**IMPORTANCE OF TEAM WORK IN VE/VM**

| | |
|---|---|
| OBJECTIVE:<br><br>To emphasise that decisions produced by group which interacts (i.e. based on consensus) in reaching a decision are superior to both the majority of individual judgements contributing to it or the average of individual decisions.<br><br>IMPORTANT:<br><br>For Senior Executives, no need to go in details (up to GAMES).<br><br>Lecture Method: (up to "GAMES"). Jot down the points from the text of "Statement" and cover the portion by lecture method in about 30 minutes for working level executives and in about 10 minutes for Sr. Executives. | <br><br><br><br><br><br><br><br><br><br><br><br><br><br><br><br><br><br>Why in groups? Why not individually?<br><br>To answer this question, we must go deep into the various aspects involved. |

| Procedure | Statement |
|---|---|
| | What aspects of a product are likely to be affected by changes that will be brought about by Value Engineering? |
| BRINGOUT: (Record in 4th column of CB)<br><br>- Engineering/Design<br>- Method/process (including tool design)<br>- Scheduling<br>- Purchasing<br>- Maintainability/Serviceability<br>- Quality requirements<br>- Costing methods<br>- Marketing/commercial aspects, etc. | |
| If answers are: | (Any change in its design? And/or method of manufacture? What about quality requirement? Any repercussion on costs?_____) |
| "Change in specifications of materials, tolerances," etc., relate them to one of the above; get agreement of the participants before recording on C.B. | |
| | Thus, a change in design normally has its ramifications in all other operating departments. |
| | It may require review of existing facilities vis-à-vis additional facilities that may be required to produce the new design product. |
| | Once the method is changed, work standards will have to be revised, and if the labour charges are based on piece- rates  or if incentive scheme in operation, some  industrial relation problems will have to be anticipated |
| | The scheduling (PPC) dept. will have to decide the exact cut-off date and carryout the economics of modifying, scraping or using "as-is" the parts on hand or making matching parts till the stocks on hand last. |

| Procedure | Statement |
|---|---|
| | Purchasing function will have to be given sufficient time for locating sources of suppliers or determining vendor prices for the proposed changes and arranging supplies in time. |
| | Quality, serviceability, warranty and also the likely reaction of the customer will have to be taken care of by Marketing and Q.C. Depts. |
| | Potential saving resulting from the VA/VE studies cannot be checked without the help of estimating or costing section. |
| | Thus, as I said earlier, the impact of a value study will have its reverberations through a major part of organisation. |
| | Who can tell the extent to which various departments are affected? |
| BRING OUT:<br><br>"People from the concerned depts." | No individual is equipped with sufficient knowledge about all the aspects of a product – its design, manufacturing , methods, costs, etc. |
| | It is too much to expect, for example, the Methods Engineer to shoulder the burden of design calculations and cost evaluations. |
| | Well, then, what should be our approach to Value Engineering study? (individual or group?) |

| Procedure | Statement |
|---|---|
| PAUSE for obvious response "group" or "team working". | |
| | The group or team approach permit pooling if available talent and potentials in the organisation for the common cause of cutting cost or improving value. |
| | It also helps us in overcoming the proverbial "resistance to change" and in facing the critical scrutiny from the line management and securing their cooperation. |
| | As you may be aware, the moment you suggest a change, however sound or logical it may be, you are likely to confronted with a barrage of questions and comments: |
| | - Why change now, when we have been doing like that for a long time? |
| | - Nobody is supplying that material. |
| | - Will there be any <u>real</u> saving? |
| | - Will the customers accept it? |
| | - And so on and so forth. |
| | Who can answer such questions satisfactorily? |
| PAUSE for anticipated response "specialists from functional areas". | |
| | Well, this answers our next obvious question as to what should be the composition of the team. |
| | Will you, then, please specify broadly the functional areas that should represent on a VM team? |

| Procedure | Statement |
|---|---|
| BRING OUT:<br><br>Representatives from<br>- Engineering<br>- Methods<br>- Production<br>- Cost<br>- Q.C., etc. | Suppose one of us wants to buy a new T.V set or a refrigerator. How does he go about selecting it? |
| BRING OUT:<br><br>- "Consults dealers" (for specifications, design, price, spares).<br>- Consults friends and relatives owning similar equipment (reg. performance, after-sale-service etc.) | (Does he go alone to shop, straightway make payment and come home with a package? or does he consult anybody before taking a final decision? IF yes, whom? why?)<br><br>In case of serious ailment, what method is adopted by hospitals in deciding whether a course of medicines should be administered or operation carried out? |
| BRINGOUT:<br><br>- Joint consultation between the specialists – or team of specialists<br>- Product Committee<br>- Board of Directors | In a manufacturing concern who takes decision on diversification or long term investment?<br><br>In matter of National/Defence strategies how are decisions made and who authorises the operations? |

| Procedure | Statement |
|---|---|
| - Joint Chiefs of Staff.<br><br>- Cabinet Committee.<br><br><br>Substance of "Statement" may be given in a 'matter – of – fact' manner, while going round the tables. | The point that stands our clearly from these examples is that all important decisions are made by groups of people – specialists in their own fields- and not by individuals, be the problem relate to a family need or national security or world affairs.<br><br>Research in the area of group-decision making indicates that individual decisions tend to be less accurate than some form of group decisions at least 51% of the time.<br><br>Also, decisions produced by group which interact in reaching a decision are superior to both the majority of individual judgements contributing to it and the average of individual decisions.<br><br>The quality of decision improves in the groups in which the members actively discuss the issues threadbare.<br><br>This is known as 'Consensus Technique' No decision becomes final which cannot meet with the approval of each and every member.<br><br>For this reason, consensus is very difficult to achieve and requires a fairly sophisticated understanding of the dynamic of conflict, interpersonal sensitivity and use of internal group power.<br><br>However, this difficulty (of obtaining consensus) can be minimised considerably. |

| Procedure | Statement |
|---|---|
| | - If seniority and 'aura' of superiority associated with seniority are dropped when the group is in sessions, thereby permitting a free and frank exchange of ideas;<br><br>- If all the functional areas are treated with respect and given equal importance; and<br><br>- above all, if the chairman or coordinator of VM. Group is not only an expert in VM. Techniques but acknowledged to be an unbiased person with no particular axe to grind.<br><br>This implies that the suggestions should always be considered impersonally i.e. from the team rather than the contributor's or departmental view point so that the final decisions are essentially the team decision. |
| CLEAN 4th column of the C.B. | |

### GAMES ON IMORTANCE OF TEAM WORK

| Procedure | Statement |
|---|---|
| - Announce Teams.<br><br>- Play at least on game.<br><br>1. WIN AS MUCH AS YOU CAN.<br>    OR<br>2. ASSEMBLE AS MANY REGULAR GEOMETRICAL FIGURES AS YOU CAN FROM THE GIVEN 64 CUBES.<br>3. GAME OF BALLOONS<br>    OR<br>4. Any other game highlighting importance of team work may be introduced. | [Refer to P.257] |

| Procedure | Statement |
|---|---|
| | Let us play a game called "_____". Has anybody played this game earlier? If so, please come over here and help me/us (if an expert is conducting the game). The game is played like this; |
| Explain how the game is played. | |
| Make sure that you have thoroughly understood the rules and 'intricacies' of the game and can comment expertly on the final 'group score'. If not, invite an expert to play this game. Also ensure that no body familiar with the game is allowed to participate now. | |
| | (Introduce the expert; |
| | Here is Mr. _____ who has kindly consented to conduct this session. Over to you, Sir). |

## ADVANTAGES OF TEAM WORK

| Procedure | Statement |
|---|---|
| | Let us look back at our discussion so far on team work and sum up its advantages. |
| Bring out: | |
| - Cross-fertilization of ideas yield sound results. | |
| - Free and Frank exchange of ideas without any inhibition encourages fresh thinking. | What are the benefits of team work? |
| | Do you recall the axiom that two heads are better than one? What does it signify? |
| | If simply means that probability of obtaining a good sense level is improved by using more people in making a decision. |

| Procedure | Statement |
|---|---|
| - Increased cooperation between depts. Creates congenial atmosphere to introduce any change.<br><br>- Appreciation of others' work. | Why seniority should be dropped when team is in session?<br><br>Why should the team include men from different discipline?<br><br>Besides these advantages, the team work in VE helps in laying stress on spreading the cost-consciousness and reappraisal of general specification without any offence.<br><br>It also promotes development of latent talents of individuals.<br><br>Last but not the least, it generates a feeling of participation in value studies, thereby permitting smooth implementation of suggestions. |
| CLOSE SESSION<br><br>Thank the expert and participants.<br><br>Clean the C.B except Column 1. | |

Very Important Note:

In case the participants come from managerial level, needing no lecture on importance of Team work and if, at the same time, the invited expert could not make it to conduct the Games, choose one of the following 3 alternatives:

1. Advance the next session.
2. Arrange a visit to WORKS/VE EXHIBITION.
3. Arrange a film/slide show on VA/VE or on Case Studies.

# 5

# VALUE ENGINEERING WORKSHOP

## SESSION - IV
## — SELECTION PHASE

# VALUE ENGINEERING WORKSHOP

## Session-IV:- SELECTION PHASE

| Procedure | Statement |
|---|---|
| **DISCUSSION OF PHASE-1: SELECTION** | |
| <u>Open Session:</u> | In this session I would like to elaborate the first phase of the job plan- called "Selection". |
| <u>C.B.(2nd Column)</u><br>Tick off "1: Selection" in 1st column. And write | |
| "1.SELECTION" in 2nd column. | There is no mathematical formula or thumb rule that will help us in identifying areas for VE study i.e. product, material, Service, or operation that seems likely to bear most fruits. |
| <u>Product life cycle</u> | |
| | Let us first look at the product and sale profile of a modern manufacturing company. |
| | In today's industrial environment, a very few firms can survive by producing a single item. |
| | The product profile of most manufacturing firms is like a joint family – consisting of some senile members, some mature persons, some fast growing youngsters and some in conceptual and gestation stages. As one product starts waning, another one or two enter in the vigour of youth to sustain rising trend in the total business of the firms. |
| | If sale volumes are projected against a time scale, the total business volume will be something like this: |
| EXHIBIT SLIDE/FC-12/IV | |

| Procedure | Statement |
|---|---|
| | Can you relate this picture to your products? Do you remember what was the first product of your division? What are the other products and when were they introduced? – Any plan for new products? |
| ALLOW brief review. | |
| | What are the products in different stages of the market life cycle? |
| No comments. | |
| | As can be seen from the market life graphs of the individual products, products are mortals like human beings – they are born and ultimately die. |
| | An individual products' life cycle may be divided into three major periods: |
| | – the product youth: |
| | – the product maturity; and |
| | – the product senility or decline. |
| | This is shown on this chart. |
| EXHIBIT SLIDE/FC-13/IV | |
| | Some products may have sky rocketing sales but short life (e.g. yo-yo and hoola – hoop) while the others may stay in the maturity period for decades (e.g. hammers, pliers, spanners and other hand tools, "detergent cake. "Vim," "Cadbury" chocolates). |
| | The duration of each period depends on several factors – approval by the pioneering triers, sales promotion, technological innovation, competitors' entry into the market, availability of newer competing products fulfilling the same needs and of superior value in buyer's view, and so on. |
| | Once the decline starts for a product, search begins for giving birth to a new product. |

| Procedure | Statement |
|---|---|
| | Vigorous R & D activities coupled with product policies and strategies sustain the company's growth of total business. |
| | Keeping in view the market life cycle, can you suggest the stages of cycle in which a product should and should not be selected for value studies? |
| The suggestive statement may bring forth the <u>obvious reply:</u> "Select a product before its decline" and not during its senile period. | |
| | If the survival is protracted over many years? – Say, because the sellers yet may find it profitable and buyers are sticking to brand loyalty and both put efforts into sustaining it. |
| "Then we should apply VM/V.E. Techniques to such products also." | |
| | Merely because the product has entered into decline phase, we should not neglect it. |
| | Of course, the new products should always be subjected to VE techniques to derive maximum profits and to increase market penetration through competitive pricing. |
| | When a potential product is 'conceived' out of knowledge, observation and inspiration into an embryo product ideas gestation begins and funds are sunk to develop the "off-spring" fully before its birth to create the necessary healthy market environment for its survival and growth. |

| Procedure | Statement |
|---|---|
| <u>C.B.</u><br><br>Add in 2<sup>nd</sup> column below "Selection".<br><br>A) *PRODUCT LIFE CYCLE<br><br>    - Conception, development, birth<br>    - Youth,<br>    - Maturity<br>    - Decline<br><br>    'New products'<br><br>B) *SALES X PROFIT ANALYSIS: | Next, sales Vs profit analysis provides a good starting point for VA/VE studies.<br><br>A systematic analysis is likely to confirm the validity of Pareto's Law.<br><br>Some of you may be familiar with this law as applied to A-B-C analysis in Inventory Control.<br><br>I will briefly touch upon it before taking up Sales Analysis.<br><br>In 1897 an Italian Economist and Mathematician Vilfredo Pareto discovered a Significant relationship between National income and population. Pareto found that a very large proportion of Income was attributed to only a small Percentage of the total population.<br><br>Later on it was found to be applicable not only to national income and population but in many other areas. The most appropriate definition is that:<br><br>    "The significant elements in a specified group usually constitute a relatively small portion of the total items in that group." |
| Select 2 or 3 examples from those given hereunder or quote from any other fields. | |

| Procedure | Statement |
|---|---|
| | <u>Some examples:</u><br><br>(1) It may be noticed that in a medium size industry stocking thousands of items of raw materials, standard parts, tools, spares, stationery, etc. about:<br><br>  – 10% of high value items account for 70% of the total value of the inventory:<br>  – 70% of low value items account for only 10% of the total inventory value, the remaining:<br>  – 20% of total value being the share of remaining 20% of the items. |
| <u>C.B. (4<sup>th</sup> column):</u> | |

C.B. (4$^{th}$ column):

| | % of items | % of value |
|---|---|---|
| A | 10% | 70% |
| B | 20% | 20% |
| C | 70% | 10% |

(2) A large aircraft may have 0,5 million parts, but 90% of its cost might be represented by only 100 or so of these parts.

(3) 90% of the company's distributors may account for only 20% of the sales, while the other 10% moves 80% of the goods.

(4) About 10% of the employees may account for 80% of the grievances.

(5) 5% of the customers makes 90% of the complaints.

(6) 10% of people in a meeting consume 80% of time.

(7) 90% of the orders comes from 5% of its clients and so on and so forth.

| Procedure | Statements |
|---|---|
| | Briefly, any group or system has: <br> **<u>"a vital few and trivial many"</u>** <u>elements.</u> <br><br> Coming back to Sales Analysis, it will reveal a similar trend i.e. some 10 to 25% of the products accounting for 80 to 90% of the profits, and 60% to 70% of Items yielding only 5 to 10% of profits. <br><br> It seems logical that out attention and efforts should be concentrated on high volume – low profit items because even a small (marginal) improvement on such items would be multiplied by the large volume sale. <br><br> Next order of priority for V.A. study falls on low volume and low profit Items. Why? |
| Get one or two illustrative examples to drive the point that the savings would not be commensurate with the expenses and efforts put in. | |
| | Low volume – high profit products, though given a low priority for VE/VM studies, should not be overlooked. <br><br> Probably, the Value Engineering study will lead to increased total profits by Increasing the Sales Volume through lower costs and pricing. <br><br> So what should be our first step in Sales Analysis? |
| Lead discussion to bring out: <br><br> "Profit factors of all The products" <br><br> *Sales volume x profit Per unit. | |
| | Having determined the profit factors, What should be our next step? |

| Procedure | Statement |
|---|---|
| BRINGOUT:<br><br>"Ranking by profit factors" | |
| | This secondary ranking is obtained by using the criteria:<br><br>(1) High Volume – Low Profits<br>(2) Low Volume – Low Profits<br>(3) Low Volume – High Profits |
| C.B.<br>(below "Sales Analysis")<br><br>1. High Vol. – Low Profits<br>2. Low Vol. – Low Profits<br>3. Low Vol. – High Profits | |
| | From our discussion on market life cycle and also from your experience of intermittent production run and investment in tooling etc., can you suggest some ways and means of refining this ranking? |
| BRING OUT:<br>Rule out:<br><br>– Items in their 'declining' stages of market life;<br>– Items with spasmodic (intermittent) production runs;<br>– Low volume items requiring high investment in design, tooling and related charges where the cost of such changes might more than off-set in marginal gains. | |
| Modify the statement to suit the composition of the group. | (In heavy engineering industry, product analysis is not always feasible because the main products are a combination of a large number of other complex units. |

| Procedure | Statement |
|---|---|
| | e.g. a turbo-generator set consists of a turbine, a generator, heat exchangers, condensers, pumps, etc., each one of them, a complex product by itself.) |
| | So is the case with heavy machines, aircrafts, etc. |
| | In such cases, the best way is to start with the analysis of the total cost components, viz: |
| |    – material cost;<br>   – operation cost;<br>   – labour cost;<br>   – over head, etc. |
| C.B. (2nd column):<br><br>C) *    COST ANALYSIS<br>    – material cost<br>    – operation cost<br>    – labour cost<br>    – overheads, etc. | |
| | This analysis helps in identifying value opportunities in various cost factors. |
| | May I remind you of the case study we discussed in 2nd session regarding the 33 KV MOCB. That comes under this category of analysis. |
| | Let us go to yet another source which provides means for selecting V.E. Projects. That source is Q.C. reports. |
| C.B. (3rd Column):<br><br>D) *Q.C. REPORTS | |
| | Normally, these reports give quality indices for: |
| |    – incoming materials;<br>   – shop-made items;<br>   – sub-contracted parts; and<br>   – castings and forgings. |

| Procedure | Statement |
|---|---|
| | In case of incoming materials, the tables of such indices contain: |
| | - No. or value of goods inspected, |
| | - No. or value of goods rejected, |
| | - No. or value of goods accepted after deviation or rework, etc. |
| C.B. (below "Q.C. REPORTS")<br><br>– High value – high rejection.<br>– High value – low rejection.<br>– Medium value – High rejection.<br>– Low value – High rejection. | |
| | In case of shop production, analysis of non-conformances (i.e. reworks, rejections, and deviations) provide detailed information and cause-wise percentages of rework and rejections, value of rework in terms of standard hours and equivalent production value, etc. |
| | Similar information but in terms of weights and values is available in appraisal reports on foundry and forge shop operations. |
| | Here are some extracts from Q.C. reports of a large engineering industry having its own captive foundry. |
| DISTRIBUTE HAND-OUT<br>VA-04/IV<br><br>Read out a few lines from each table. Allow 10 minutes to go through the reports. | |
| | If you were to select a project based on these reports, which area will you attack first? why? |

| Procedure | Statement |
|---|---|
| | Well, let us carry out a small exercise based on these reports. |
| Preferably same teams as formed at the end of SESSION-II. | For this purpose, we will work in groups, and I suggest the composition of Groups like this: |
| | Group-I |
| – You might have spotted 'potential leaders' by this time.<br><br>   Include at least one such 'leader' in each group.<br><br>– (Repeat till all participants are covered)<br><br>– Keep group strength limited to 4 or 5 members.<br><br>– Explain sitting arrangement. Time allowed and the nature of assignment i.e. analysis.<br><br>– Allow 10 to 15 mts. | Mr._____<br>Mr._____    } Will form the first group<br>Mr._____ |
| When re-assembled: | |
| One-by-one: | |
| ask the leaders of the groups to briefly explain their respective cases. | |
| Summarise and put down the points on C.B. | |
| (4th column) | |
| Group          Points<br>                     (reasons) | |
| I          . . . . . . .<br>II          . . . . . . .<br>III          . . . . . . . | |

| Procedure | Statements |
|---|---|
| In case there is no common agreement, the group leaders themselves form a group and arrive at a consensus decision.<br><br>– Allow 5 to 7 minutes.<br>– Record final decision on CB.<br>– Allot rank nos. and PUSH ON. (No further comments required). | |
| | Similar approach can be adopted in case of "In-service" or "Field-service" reports. |
| <u>C.B. (3<sup>rd</sup> column):</u><br><br>E) *    <u>FIELD SERVICE REPORTS</u><br><br>If "QC reports" part is fully discussed, there is no need to go in details of Field Service. reports. | |
| | Perennial scarcity of certain materials, excessive dependence on imports of know-how and machinery, sky-rocking escalation in prices, etc. offer great challenges and also opportunities to Value Engineers in selection of V.E. studies. |
| <u>C.B. (3<sup>rd</sup> column):</u><br><br>F) *    <u>SCARCE MATERIALS</u><br><br>G) *    <u>IMPORT SUBSTITUTION</u> | |
| | An increase in efficiency of kerosene stove by 2% is estimated to save about Rs. 500 millions in Foreign Exchange annually (based on 1980 prices).<br><br>In a country like ours – blessed with bright sun-shine through most part of the year, a saving of 5% on conventional energy sources for heating or lighting will show a saving of staggering amount to the nation. |

| Procedure | Statement |
|---|---|
| | Lastly, no method is better than discussion method between knowledgeable people in selection of a V.E. project. |
| C.B. (3<sup>rd</sup> column):<br><br>H) *   <u>DISCUSSION BETWEEN KNOWLEDGEABLE PEOPLE</u>. | |

### JOBS FOR VE/VM WORKSHOP

| Procedure | Statement |
|---|---|
| BRIEF MEMBERS FOR NEXT SESSION. | Now a word or two about jobs to be selected for our "workshop".

This is where I have to appeal for your help and cooperation in providing jobs suitable for value analysis in the class.

You might have already thought of such jobs or your seniors might have selected the jobs for our practice in the class.

If yes, I would like to discuss with you after a while; if not, please discuss with seniors in your department and look for jobs that can be described in the class in about 10 minutes.

It may be:

- a small assembly;
- a measuring or protective device;
- a wrapping and packing job;
- perhaps the one which is giving you trouble either in service or in assembly or in procurement.

Whatever be the job you select, let it be one from your own department or one about which you have reasonably sound knowledge.

Please bring with you the drawings, samples, process sheet, small hand tools (if required for assembly), and any other data required for demonstration. |

| Procedure | Statement |
|---|---|
| | If the job relates to formats/form, or any other kind of paper work, it would be advisable to bring sufficient copies so that each participant could have a copy to look at. |
| | It is very important that the whole job can be <u>demonstrated</u> in this room. |
| | No riddles or puzzles, please. |
| | I will see each of you before your turn comes for demonstration in later session so that you and I will be clear as to what is required and what the problem is. |
| | If you have already selected the Job(s), please come to the head of the table with necessary drawings and data. |
| Invite 2 or 3 participants one by one; make them describe the jobs and point out their suitability for class work.<br><br>Clear doubts. | |
| | Tomorrow I will compile a list to ensure that we have different types of jobs to provide greater interest. |
| | Here is your opportunity to dream up all possible solutions to these problems. |
| | Forget quality; get a large number of ideas. jot them down <u>as and when they occur</u>. We will need them in our future sessions. |
| CLEAN BOARD<br>Retaining the 1<sup>st</sup> column.<br><br>CLOSE SESSION with suitable statement. | |

# 6

# VALUE ENGINEERING WORKSHOP

SESSION - V

— INFORMATION PHASE-I

(DATA COLLECTION)

# VALUE ENGINEERING WORKSHOP

## Session-V:- INFORMATION PHASE-I

## (Data collection)

| Procedure | Statement |
|---|---|
| Ensure group-wise sitting arrangement<br><br>OPEN SESSION | |
| **REVIEW OF PREVIOUS SESSION** ||
| <br><br><br><br><br><br><br><br><br><br><br>* State only if the game was played<br><br><br><br><br><br><br><br><br><br>C.B.<br>Put a trick mark against item 2 in column 1 and write in 2<sup>nd</sup> column<br><br>"2. INFORMATION" | In the first session we discussed the concept of value. Then we consider the causes of unnecessary costs that creep into the product<br><br>In the 2<sup>nd</sup> session we identified the several phases of Job Plan<br><br>After that we agreed on the importance of consensus reached in talking a decision.<br><br>*(The importance of team work was high-lighted with the help of game: "win as much as you can" OR GAME OF BALLOONS Page 246)<br><br>In the last session, several methods of selection of a project for VE/VM Study were discussed.<br><br>Now we enter into a crucial phase of Value Engineering, namely, INFORMATION or Data Collection Phase. |

| Procedure | Statement |
| --- | --- |
| **SOURCES AND VALIDITY OF INFORMATION** | |
| Objective:<br><br>To put stress on collecting information from right sources in right amount and at right stages. | The actual exercise in VE/EM starts with collection of information relevant to the selected product or subject, and it presents a real challenge. |
| | Only if you seek the required information from the authentic sources will you succeed in your VA/VE exercises. |
| | Once all the facts are available there is an excellent chance that the problem will find its own solution. |
| | Before discussing what information is to be obtained and from which sources, let us spend few minutes in looking at the flow of materials and documents through an average sized manufacturing organization. |
| Confine board work to R.H. side.<br><br>  – Open discussion with appropriate questions depending on the type of industry.<br><br>  – A suggested questions are given in Statement Column.<br><br>  – Certain questions like "who decides the batch size?" will not be applicable to process or job industry and hence, should be avoided if the participants come from such industry. | |

| Procedure | Statement |
|---|---|
| **FLOW OF INFORMATION IN A MFG. INDUSTRY** ||
| | Can anybody help me in tracing the flow of documents and materials? |
| DISTRIBUTE COPIES OF DOCUMENTS FLOW CHART--PAGE 95 | |
| Encourage narration of flow. Allow brief discussion, sum up in the form of a Document Flow diagram (as given at the end of this session's notes – P.98) | |
| DISTRIBUTE COPIES OF 'DOCUMENTS FLOW CHART'. | |
| If response is not encouraging, put the suggested questions. | What is the main objective of a manufacturing company? |
| If the answer is "Make profit," put the next question to bring out: | - How do they make profits? |
| "By making and selling goods or providing Service." | - Think of major departments, their functions and sequence of flow of information through them. |
| Encourage 1 or 2 participants to come to head of table and to draw 'Information flow' diagram. | |
| Use the remaining questions as a guide. | |

| Procedure | Statement |
|---|---|
| - Marketing.<br><br>- Engineering.<br><br>- Methods.<br><br>- Purchase.<br><br>- Technology/process<br><br>- Scheduling/PPC(Production Planning & Control)<br>- Commercial/PPC<br><br>- Q.C/Q.A.<br><br>- Costing/Finance<br><br>- Shipping/Dispatch. | - Who knows the market demand?<br><br>- Who books orders?<br><br>- Who prepares detailed specifications?<br><br>- Manufacturing method?<br><br>- Why purchase buys materials of certain specifications and not as it likes?<br><br>- Who authorises drawal of materials from the Stores? And why a fixed quantity?<br><br>- Who decides the priority of manufacturing, batch size, etc.?<br><br>- Who checks and certifies conformance to quality standards?<br><br>- Who can tell us whether we are making profits or losses?<br><br>- Who takes care of delivering the goods to customers? |

| Procedure | Statement |
|---|---|
| Write down the names of functional areas under "INFORMATION" in 2<sup>nd</sup> Column. <br><br> - Don't be tempted to give your own views either on nomenclature used for denoting the functional areas or the systems & procedures followed in different organizations. <br><br> This may involve you in arguments and cross talks or faulty reasoning. <br><br> Both are 'time killers'. <br><br> The 2<sup>nd</sup> column board work will appear thus: <br><br> (Use Handout VA-O5/V) <br><br> For bringing out sub-heads under each of the headings. No need of writing sub-heads on C.B) <br><br> "2:INFORMATION" <br><br> * Marketing/Commercial <br><br> * Design/Engineering <br><br> * Scheduling/PPC <br><br> * Methods/Technology <br><br> * Materials Management/Purchase & Stores <br><br> * Manufacturing/Production <br><br> * Q.C/Q.A./Inspection <br><br> * Finance/Costing/Accounts <br><br> * Shipping/Dispatch | <br><br><br><br><br><br><br><br><br><br><br><br><br><br><br><br> What sort of information would you expect from these departments? |

| Procedure | Statement |
|---|---|
| | Let us start with the 'Marketing' |
| If no appropriate answers come forth for, put suggestive questions to bring out the necessary information given under each starred title. | |
| * <u>MARKETING</u> | |
|   - Customers requirements like reliability, serviceability, warrantee. | |
|   - Anticipated market demand. | |
|   - Application. | |
| | Next, Design Department. What Sort of information is sought for from it? |
| * <u>ENGINEERING/DESIGN:</u> | |
|   - Dimensioned drawings of parts, assemblies. | |
|   - Latest specifications. | |
|   - Weights of parts. | |
|   - Tolerances. | |
|   - Operating conditions like max. & min. temperatures Pressure and other environmental factors. | |
|   - ………. | |
| Cover the remaining departments one by one in like manner. | |
| * <u>SCHEDULING/PPC:</u> | |
|   - Operational planning sheets. | |
|   - Scheduling. | |
|   - Capacity Vs. Make or buy decision. | |
|   - ………….. | |

| Procedure | Statement |
|---|---|
| * METHODS:<br><br>  - Process specifications (machine tool, J & F, Heat & surface treatments, etc.)<br><br>  - Material preparation plans (Material utilization, cutting allowances, tolerances, etc.)<br><br>  - Operation time.<br><br>  - Level of skills involved.<br><br>* MATERIALS MANAGEMENT:<br><br>  - Suppliers (current & potential)<br><br>  - Vendors performance.<br><br>  - Purchasing cost.<br><br>  - Delivery.<br><br>  - Inventory carrying cost.<br><br>  - Min. & Max. stocks & anticipated arrivals.<br><br>  - Storage, handling, distribution, disposal.<br><br>  - Scrap records.<br><br>  - Cost of bought out & sub contracted items.<br><br>* MANUFATURING:<br><br>  - Process constraints.<br><br>  - Most likely delivery dates.<br><br>  - Deviation and Re-work.<br><br>  - Available skills.<br><br>  - Models of parts or a complete set of components (dismantled)<br><br>  - Incidence of Rejection and Re-work. | |

| Procedure | Statement |
|---|---|
| **\*  Q.C./TESTING:**<br><br>- Incidence of rejection and re- work.<br>- Third party quality requirements.<br>- Inspection and Testing procedures.<br>- Limits of deviation allowed.<br><br>**\*  COSTING/FINANCE:**<br><br>- Product cost in terms of: materials, labour, overhead.<br>- Cost of rework & rejections.<br>- (Full analysis of works costs).<br>- Added Value/conversion cost.<br>- Return on investment, etc.<br><br>EXHIBIT FC-14/V;<br>Allow 2 Mnts. to study     OPTIONAL<br>EXHIBIT FC-15/V:<br><br>**\*  PACKING & DISPATCH:**<br><br>- Packaging.<br>- Handling.<br>- Shipping/Transportation.<br>- Pre-dispatch storing.<br>- Warehousing.<br><br>**\*  MISCELLENEOUS:**<br><br>- Maintenance requirement.<br>- Safety.<br>- Performance data | [As these graphs indicate, saving potentials are maximum & implementation cost, minimum, at the beginning of the project or product life. The net gains diminish from 50% to 5% when the product gets stabilized.] |

| Procedure | Statement |
|---|---|
| - Consumption pattern.<br><br>- Environmental influence.<br><br>- Difficulties in storage, handling, disposal, shelf-life, etc.<br><br><br><br><br>*(HARDENED GLASS LAMINATE)<br><br><br><br><br><br><br><br><br><br><br><br><br><br><br><br><br>DISTRIBUTE HANDOUT VA-05/V<br><br>Talk while going round the table. | e.g.<br>(1. Cutting oil causes dermatitis:<br><br>- Should be non-corrosive to job & machine tools.<br><br>- Should be non-toxic and non-irritant.<br><br>- Should be stable during service life.<br><br>- Should be compatible with m/c lubricants.<br><br>2. *HGL dust and fumes of acids, paints, thinners and other chemicals are injurious to lungs and eyes.<br><br>3. In surface treatment shop, many solutions require neutralisation of effluent before disposal.)<br><br><br>Here is a handout listing the information that may be sought for from various agencies.<br><br><br><br>The list is neither extensive, nor exhaustive. Requirements of information will vary from project to project and has to be obtained in a 'Tailor-made' fashion.<br><br>However, it is worthwhile to repeat the caution.<br><br>"Seek information from the most authentic source only" – and not from the so-called knowledgeable persons because the information given by such persons may be based on insufficient knowledge, heresy, guess work, biased mind or preconceived notions. |

| Procedure | Statement |
|---|---|
| | (Physics or Chemistry does not work as we think – it works as per a set of Laws). |
| | Such information will not only mislead and confuse you but is likely to land you in trouble in the future phases of the VE programmes. |
| | Can someone pass on his own experience of this sort? (That is, information given by a so-called knowledgeable person has put you in embarrassment.) |
| Get 2/3 examples based on participants' own experience. (See page 96) for author's experience). Offer no further comments. | |
| | The creative phase actually starts in this phase itself, i.e. in the phase of collecting facts. |
| | There is disagreement about the extent to which one should go in gathering the Facts. Some creativists believe on saturating with facts and more facts. |
| | Some fear that too many facts at the wrong stages in creative problem solving may swamp our imagination and lull us into a willingness to accept the obvious solution and thus shut us off from thinking up anything really new. |
| | The best course is to start by lining up a few salient or fundamental facts related to the problem, which are many a time more helpful than the facts directly in point and really at hand. |
| | A conscious search for material beyond obvious will also help in formulation of all possible hypotheses. |
| | Cause and effect are usually the most important facts to find, because they clarify our thinking and help us in focussing on the problem objectively. |

| Procedure | Statement |
|---|---|
| | The facts we gather and the data supplied by memory and observation should help us in preparing a frame work of ideas that, in turn, will serve best to further our creative thinking. |
| Clean C.B. except 1<sup>st</sup> column<br><br>Distribute Hand outs VM-04/VII,<br>            VM-05/VII<br>            VM-06/VII<br>            VM-07/VII  and<br>            VM-08/VII | |

# DOCUMENTS FLOW CHART

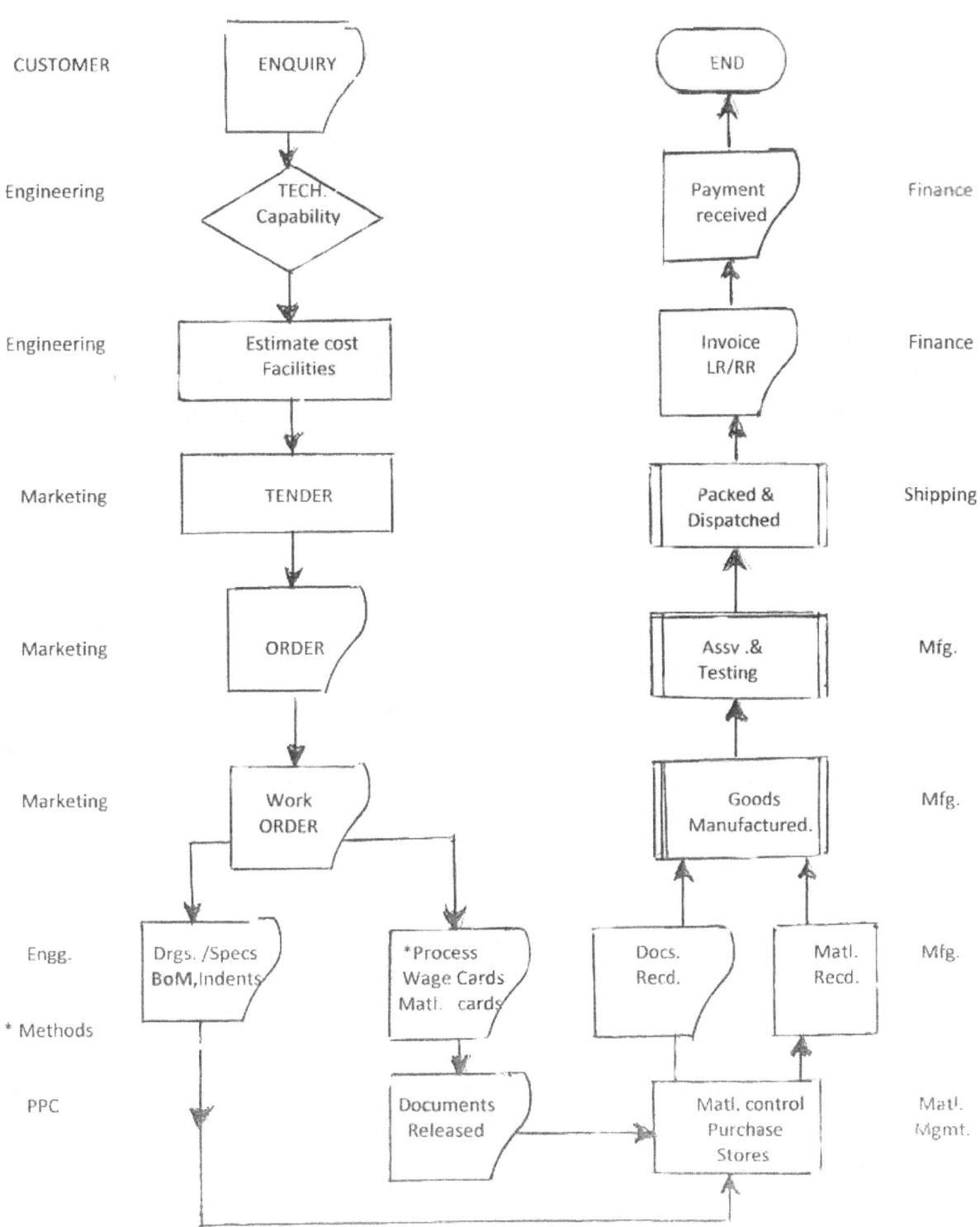

## EXAMPLES FOR SESSION-V

Ex.1    On my first visit to Bangalore, I told a fellow bus passenger who was standing next to me of my destination and requested him to alert me at one stop earlier. He did it promptly and I thanked him profusely. I got down from bus — only to discover, to my great dismay, that my destination was miles away! On enquiry at that odd bus stop, it became clear that he had played a trick on me to occupy my seat!

Lesson : I should have got the information from the conductor, who was the right source of it.

Ex.2    Based on the drawings and specification furnished by the design office, a VE team selected heavy aluminium casting cable cover for VE study and after taking it through the full drill of VE phases, came out with the suggestion that it could be replaced with simple sheet metal piece bent to shape. On going through the VE report, the chief designer called me (as the VE Coordinator) and rebuked me and the team for reinventing the wheel" and wasting company time.

It was later on learnt that on suggestion of a worker, the aluminium casting was replaced by a pressed sheet about 6 months earlier. The team member instead of approaching the design, had collected the information from the technician who was not aware of revised drawing and specifications and had given the old set of documents to the team leader.

# 7

# VALUE ENGINEERING WORKSHOP

## SESSION - VI

## — INFORMATION PHASE-II

### (Parts, Function & Cost)

# VALUE ENGINEERING WORKSHOP

## Session-VI - INFORMATION PHASE- II

## (Parts, Function & Cost)

| Procedure | Statement |
|---|---|
| **DEFINITION OF PART & ITS FUNCTION** ||
| Open Session with suitable remarks.<br><br>C.B. (3rd Column)<br><br>PART & ITS FUNCTION(S): | The second part of 2nd phase relate to parts and their functions.<br><br>The next step in our deliberation is to define the part and its functions. This is a very crucial piece of information- crucial because in VE/VM we do not attempt to alter or eliminate function.<br><br>Rather we attempt to provide the function at least cost. |
| C.B | |
| 1. What is it? | What is it?<br><br>i.e. What is this part or product?<br><br>A quality answer to brings out a precise definition of the part or the product.<br><br>That is important because names often misleading and convey wrong impression about the functions performed or purpose served. Take for example, |
| Mention 2 or 3 'names' with appropriate comments. | (1. Hand shearing Press- does it shear hands?<br><br>2. Cobra Saw - has no cutting teeth.<br><br>3. Monkey Spanner- God alone knows the origin of this fancy name<br><br>4. Lathe dog – for pawl.<br><br>5. Milk cooker – have you tried cooking milk ? |

| Procedure | Statement |
|---|---|
| | Well, that is a part of Indian English, I suppose.<br><br>6. Butter milk – There is neither butter nor milk in that!<br><br>7. Dog house (in Oil Rigs)<br><br>8. Cat walk – and so on)<br><br>May I request you to contribute to this list of misnomers? |
| Allow few minutes.<br><br>Offer no comments.<br><br>Watch out & discourage names of persons:<br><br>e.g. "Dhanaraj" or "Dhanalaxmi" or "Koteswara" – begging for a few coins. May be there is a participant by that name or his near relative bears that name and he is sure to fell offended.<br><br>C.B (4<sup>th</sup> column) | |
| | Here is one more examples :<br><br>From the component drawings, I have noted it being variously called as |
| | - Bush with collar<br>- Bush bearing<br>- Flanged bush<br>- Distant element<br>- Spacer<br>- Catch Spacer<br>- Washer (thrust)<br>- Stepped round Washer<br>- Sealing Washer<br>- Contact Washer<br>- Plain bushing.<br><br>From its name, can you suggest a single purpose served or function (s) performed by it?<br><br>Now, we know that a bush, flange, spacer and washer have different function and a single part may not be able to perform all the functions. |
| Pause:<br><br>LONG PAUSE: | |

| Procedure | Statement |
|---|---|
| | Equally important is the definition of the function or functions performed by the part or the product. |
| C.B. (3rd Column) | |
| 2. What does it do? (key procedure) | What does it do? |
| | In answer to this question lies the Key procedure of VE/VM. |
| | Its other complementary questions that must also be answered are: |
| | – What is its purpose? |
| | – What makes it work? |
| | – What makes it sell? |
| C.B. (3rd Column): | |
| Basic function: | |
| – What is its purpose? | |
| – What makes it work? | |
| – What makes it sell? | |
| **CLASSIFICATION OF FUNCTIONS** | |
| | May I remind you that a function is that attribute of a product which enables it to perform the requirements placed on it. |
| | What requirements? |
| | Let us see how and why a product is made. |
| | A logical sequence is somewhat like this : |
| | - A person (customer) becomes aware of some need to be fulfilled or purpose achieved or recognizes a function he would like to have performed. |
| | - The industry, which is always in search of newer needs in order to fulfil its own function (of making profit), spots the customer need, and develops an item to satisfy it. |

| Procedure | Statement |
|---|---|
| | - During the development stages, the industry will certainly add certain features to the product which in their opinion will attract more customers i.e. increase the saleability of the product.<br><br>- These additional features are not related to the identified need or function i.e. use value.<br><br>- They may add to the esteem value.<br><br>For example, chrome plating of faucet (tap) and door handles, anodizing of aluminium fixtures, floral designs on crockery, shining metallic strips on suit-cases, laces on petticoats, etc. add nothing to use value – they enhance sales appeal.<br><br>A product is thus born that not only achieves the basic function for which it was specially designed but also creates desire to possess it.<br><br>Thus, almost every product or service will have:<br><br>- a primary function related to USE VALUE and a secondary function related to ESTEEM VALUE.<br><br>Of course, sometimes there will be tertiary and quaternary functions and even unnecessary functions associated with a product or operation.<br><br>We will group them all under "auxiliary functions". |
| Chalk Board (3rd col.):<br><br>3. CLASSIFY FUNCTION:<br>- Primary function(s)<br>- Secondary function(s)<br>- Auxiliary function(s)<br>- Unnecessary function(s) | Lastly, there are some superfluous or unnecessary functions added to a product during various stages of its evolution. |

| Procedure | Statement |
|---|---|
| | Let us spend the next few minutes in distinguishing between different types of function. |
| | First, we will discuss <u>primary</u> functions of a few common things. |
| | Please, remember that a primary function is that which a product is specially designed to fulfil. |
| | Now, then, what is the primary function of a, say, |
| Bring out & record in 4<sup>th</sup> col. of C.B. | |
| | - Pen cap? |
| | - Cup? |
| - pen cap (protect nib or ball) | - Veena or sitar? |
| - cup (provide means of holding tea or coffee) | - Shaft? |
| - Veena or Sitar (produce/create melodious sound) | - Castor? |
| - Shaft (transmit torque) | |
| - Castor (reduce fiction) | |
| Write down the answers on Board after summing up. | |
| The choice of words should have the consent of contributor. | |
| Don't labour to make them very concise at this stage. Use of 'a verb and a noun' discipline comes later on. | |
| Watch out for secondary and tertiary functions being spelt out at this Stage. | |
| Select only primary ones at this stage | |
| | Now for the secondary functions. There are also performed by the product but if deleted would not prevent the primary function being achieved. |

| Procedure | Statement |
|---|---|
| | For example, an anodized lunch box works no way better than an ordinary box, but The beautiful/colourful finish enhances Sales appeal. |
| | The shining 'Trim-strips' on the sides of car improve its steam-lines finish but no way help in the primary function of the car. |
| | The rating specification plate on an electric motor provides certain information; motor would work even if the plate is not provided but that would be a bad sales feature. |
| | So is the case with labels on note books and trade marks on soaps and so on. |
| | The secondary functions can be identified by answering the questions:- |
| | - What else can it do? |
| | - How does it support the basic functions? |
| | - What makes it work better? |
| | - What makes it sell better? |
| <u>C.B.</u><br><br>(Draw an arrow from 'secondary function' and write below)<br><br>- What else can it do?<br>- How does it support the basic function?<br>- What makes it work better?<br>- What makes it sell better? | |
| | What features or parts of a cushioned chair perform secondary function? Its primary function is to support weight. |
| <u>Bring out:</u><br>- Cushion provides comfort.<br>- Arms support arms (provide comfort to arms) | |

| Procedure | Statement |
|---|---|
| - Rubber bushes reduce friction or noise when dragged.<br><br>- Paint prevents corrosion or prolongs service life, gives smart appearance, etc.<br><br>- Upholstery protects foam, gives pleasant feeling. | |
| | As you see, these functions are sub-ordinate to the primary function of a chair – which is to support body weight.<br><br>Can you think of some more objects and tell me their primary and secondary functions? |
| Get about 5 examples.<br><br>Sum up statements, separating primary functions from secondary ones.<br><br>(If response is poor, provide leads by mentioning the following examples:<br><br>- Washer in a tap (faucet).<br><br>- Ear of a cup.<br><br>- Collar of a shirt.<br><br>- Pelmet of curtain.<br><br>- Thermostat on iron press.<br><br>- Coiled cord for telephone.<br><br>- Chrome plating on handle bar of a cycle. | |
| | Coming to the 'useless' functions i.e. the functions which have no use or esteem value:<br><br>Such functions some-how creep into the product or are provided under wrong notions. For example.<br><br>- Humming noise in transformers or chokes of a tube light.<br><br>- Heat produced by electric lamps and electrical machines.<br><br>- Chrome plating of jotter pen refills.<br><br>- Laces or frills on under garments |

| Procedure | Statement |
|---|---|
| | Sometimes certain superfluous functions result from excessive complicated design specification, unreasonably high demands on performance such as convenience (which are not absolutely necessary), excessive safety measures or due to the whims of designer. |
| | Did you notice that some window in your house is never operated? or some door rarely closed? - Why provide them then? At least, you can avoid shutters and use panes or curtains. |
| | What is the function of expanded metal screens on windows fitted with metal bars? |
| Get a few more examples.<br><br>Discuss briefly to make the distinction between 3 classes of function clear. | |

## SPECIFY FUNCTION

| Procedure | Statement |
|---|---|
| | Having qualified the functions as being primary and secondary, we must now know how to specify the function, i.e. to describe it concisely. |
| | This is done by using two words –a Verb and a Noun - in that order. |
| C.B. (3<sup>rd</sup> column)<br><br>4. SPECIFY FUNCTION:<br><br>A VERB and a NOUN method.<br><br>State slowly and write in 4<sup>th</sup> column.<br><br>- Fire provides heat.<br>- Telephone facilitate communication.<br>- Spanner provides leverage. | - Fire <u>provides</u> <u>heat.</u><br>- Telephone <u>facilities</u> <u>communications</u><br>- Spanner <u>provides</u> <u>leverage</u> |

| Procedure | Statement |
|---|---|
| **THIS PORTION (UPTO FUNCTION OF FAN) IS OPTIONAL** | |
| Make distinction between function, purpose, and method (clear) by giving 1 or 2 examples.<br><br>(These examples are provided in statement column). | Before taking up additional examples, it is worthwhile to distinguish between the FUNCTION, PURPOSE & METHOD<br><br>This can be achieved easily if we remember the definition of the function.-"<u>Function</u> is that which a <u>product</u> or <u>service</u> does in order to achieve a <u>purpose.</u><br><br>For example, in case of an electric oven, if 'make biscuits' is the <u>Purpose,</u> 'heat or bake ingredients' will be the <u>Function</u> & use electric oven, the <u>Method</u><br><br>Similarly, <u>Fire bricks lining (Method)</u> in furnaces serve the (purpose) of <u>conserving heat</u> by preventing <u>heat- losses</u>. (function) |
| <u>C.B. (last column):</u><br><br>Record ANY ONE of the following two:<br>1.- Make biscuit  (purpose)<br>  - Heat ingredient  (function)<br>  - Ele. oven (method)<br>2.- Conserve heat (or maintain temp. (purpose)<br>  - Prevent heat losses (function)<br>  - Fire brick lining (method). | |
| Ask each group to give at least one example. | Can you think of some more example of this type?<br><br>For this purpose, we will work in groups – same groups as formed  in earlier Sessions |

| Procedure | Statement |
|---|---|
| | Each group will select a product of its own choice and briefly describe the **purpose, function and method** of the selected product – in that order. |
| <u>Allow 5 minutes.</u><br><br>Call on the 1ˢᵗ group<br>Leader to define the product, its purpose, function and method. Note down on C.B.<br>(4ᵗʰ column).<br><br>   – Get agreement from other groups.<br>   – Repeat for the other groups. | |
| | <u>FUNCTION OF A FAN</u><br><br>Coming back to concise way of describing the function, by using a **Verb** and a **Noun**, let us carryout some exercises. |
| <u>Bring out:</u><br>1. Fan – circulate air.<br>2. Flatten dough or Roll Chapatis.<br>(Some interesting definitions may emerge like 'beat husband', 'symbolise women power' etc, Accept without comments. At the most you may enliven the discussion by classifying them as primary, secondary, etc.) | 1.   What is the function of a fan?<br>2.   Function of a rolling pin? |
| 3. Field coil: Create flux. | What about the field coil in electric machines? |
|    – The expected first answer would be "write". Ask which is noun in the answer. | Writing pen? |

| Procedure | Statement |
|---|---|
| The response is "on paper," disallow 'on' and read out "write paper". Ask group whether it conveys sense. Then write on your palm or duster back and ask whether 'paper' was involved in your writing. | |
| Next draw a figure of tumbler or a spanner or a bird (whatever figure you can draw quickly) and also put a few dots and dashes in some pattern and ask whether any "writing" was involved. | |
| Idea is to draw in all members into a lively discussion. | |
| Tactfully lead discussion to the conclusion 'Make Marks' and follow similar procedure in other examples also. | |
| C.B. (4<sup>th</sup> column): | |
| 4. Pen: Makes Marks. | |
| | Why do they exhibit a "Scull & Bones" sign on H.T. poles or sub-station? |
| 5. Scull X bones: Give warning. | |
| | What is the function of a gasket or 'O' ring? |
| 6. Gaskets: Provide sealing. | |
| | Valves? |
| 7. Valve: Rregulate flow. | |
| | Switchgear? |
| 8. S.G.: Break circuit or interrupt current | |

| Procedure | Statement |
|---|---|
| If space in 4th column is not sufficient, write in lower half of 1st column. | |
| | Let us take some examples of multifunction products. |
| | What are the functions of a power tiller? |
| 9.  POWER TILLER:<br>   – Carry rider<br>   – Provide power<br>   – Plough field | |
| |   – of Chrome plating? |
| 10. Chrome plating:<br>   – Protect surface<br>   – Enhance appearance<br>   – Reduce fiction | |
| | What about a pure silk saree costing over Rs. 10,000/- |
| 11. Saree:<br>   – cover body<br>   – enhance prestige.<br>   – enhance beauty<br>   – spite neighbours.<br>     (humiliate neighbours).<br>   – please husband.<br>   – maintain domestic peace/bliss.<br>   – indicate festivity.<br>   – attract attention.<br>   – exhibit affluence.<br>   – please oneself. | |
| 12. Diamond Ring. | |
| | Diamond studded gold ring? |
| Distribute handout<br>VA-06/VI. Verbs & Nouns-<br>LONG PAUSE | |

| Procedure | Statement |
|---|---|
| | Now that we have acquired reasonably good skill in describing the function by a verb and a noun, let me mention that many a time we have to sacrifice brevity for clarity. |
| | This is done by quantifying additional requirements thus: |
| C.B. (3rd column): | |
| 5.QUANTIFY ADDITIONAL EQUIREMENTS: | |
| | In case of ex.5 about the danger sign, we may add – "be readable at a distance of 5 metres". |
| Add to ex.5 "-from 5 metres". | |
| | Similarly, the function of a gasket may be quantified by saying – "without de-formation at 450° C". |
| Add: "up to 450ºC" against ex.6 above. | |
| | In case of an amplifier, the function may be described as "amplify sound 5 times without distortion," and so on. |
| | For the next few minutes, we will practice this sort of exercise. |
| | From the Verbs and Nouns listed on the handout (VA-06/part-A), from suitable combinations to indicate functions, identify the object or item performing that function, and then quantify additional requirements. |
| | As before, we will work in groups, and each group to please provide at least one example. |
|   – Allow 5-10 minutes.<br>  – Get 4-5 examples.<br>  – Trim the description- to suit methodology. | |
| CLEAN BOARD (except upper half of col.1)<br>(SEE NOTES ON P.120) | |

| Procedure | Statement |
|---|---|
| **FUNCTION-COST ANALYSIS** ||
|| The next step in our analysis is to assess the degree of "Value Opportunity" which exists. |
|| Here, the value opportunity means the different between the actual cost incurred in providing a function and cost worth of that function. |
| C.B. (2^{nd} column): <br><br> Value Opportunity <br> = Cost (actual) – worth (estimated). ||
|| The procedure for comparing the costs of functions with their estimated cost worth, therefore, consists of <br> – analysing the function/features; <br> – determining the cost required to produce each functional feature; <br> – allot a 'cost worth' to each function; <br> and finally compare the cost to worth to assess value opportunity. |
|| This step in job plan is known as '**Cost-Function Analysis**'. |
| C.B. (2^{nd} column): <br><br> EVALUATION OF FUNCTION <br> (FUNCTION-COST ANALYSIS) <br> – What is its cost? <br> – What is its worth? ||
|| Let us first devote some time in analysing functions of 2 or 3 products and their components and noting down the same in tubular form. |
|| Here is an example of a simple item-Ink pot – as analysed by the family members of a friend (of mine). |

| Procedure | Statement |
|---|---|
| Distribute VA-07/VI:<br><br>Allow 10 minutes to study.<br><br>No comments.<br><br>Next, present a 2-pin plug, open it slowly, mentioning its various parts.<br><br>Distribute VA-08/VI:<br><br>    – Allow 5 minutes to copy.<br><br>    – Help in identifying different parts and their features. List them in col.2.<br><br>    – Discuss classes of function and record them under appropriate column.<br><br>      (Defer cost factors for few moments) | |
| | Here is an example of detailed Functional Analysis of a Transport Vehicle: |
| Handover VM-07/VI:<br>(Function levels of a Transport Vehicle).<br><br>    – Allow 10 mts. to go through | |
| | Well, gentlemen, that is not the ultimate analysis. Each part or sub-assembly of the vehicle can be subjected to a similar detailed analysis.<br><br>All that we should remember from this example is that whatever be the product, function level analysis will involve:<br><br>    – function of the product;<br><br>    – function of its sub-assemblies;<br><br>    – function of components of sub-assy;<br><br>    – function of parts of components;<br><br>    – function of features of parts.<br><br>These levels are summarised on this sheet. |

| Procedure | Statement |
|---|---|
| <u>Distribute VA-09/VI:</u><br><br>(Functional levels)<br><br><br><br><br><br><br><br><br><br><br><br><br><br><br><br>Refer to your copy i.e.<br><br>VA-08-A/VI and insert these values in col. 7 of the function cost analysis table.<br><br><br><br><br><br>Item by item, allocate or workout costs, after discussing and getting consensus. Note down the costs in last but one row. | Let us now go back to the examples of 2-pin plug and work out the cost of providing various functions.<br><br>A local firm manufacturing this type of Plugs has furnished the cost data as:<br><br>                                 (INR)<br><br>Brass pins (2 Nos.)     52<br><br>Connectors (2 Nos.)    16<br><br>Base (Bakelite)        08<br><br>Cap (      ″     )       11<br><br>Labour                <u>03</u><br><br>          Total   <u>90</u><br><br>The labour charges are based on Rs.25.0/Hr(?) and O.H. @ Rs.50.0/Hr(?). The Bakelite parts are bought out items and brass parts are shop-made using automats.<br><br>Next, let us see the costs of different functions.<br><br>Lastly, let us look at the 'cost worth' of the functions.<br><br>There are various methods of deriving and establishing function worth.<br><br>1. Comparison with <u>known costs</u> of performing the <u>same</u> function in other devices or mechanism.<br>2. Comparison with <u>known costs</u> of performing <u>similar</u> functions. |

| Procedure | Statement |
|---|---|
| | 3. Comparison with the <u>estimated</u> costs of performing the functions by the <u>simplest means.</u> |
| | 4. Apportioning a percentage of the total worth of the whole project. |
| | 5. Comparison with <u>competitors' selling prices</u> for parts which perform the <u>same</u> function. |
| | 6. Allocation of an arbitrary worth to a function, etc. |
| Ensure a lively discussion by taking up some interesting examples like that of the : <br><br> – Saree; <br><br> – Diamond ring; <br><br> – Luggage bag (suitcase); <br><br> – and similar product. | |
| | Do you remember that the primary function of a "saree" was identified as 'cover body'. |
| | Now either by comparing with known costs of covering body by other type of saree or allotting an arbitrary worth, can you tell me worth in Rupees of that function? |
| Discuss and bring out: <br> (say, Rs.125/-) | |
| | While covering the body, we would like to ensure reasonable comfort – not a coarse and rough cloth like gunny bag or a coir mat. |
| You may take the example of car door, thick Kashmiri carpet in drawing room, etc. | |
| | Let us consider the luggage boot of a car – popularly known as "dicky". |
| | What is the actual cost of providing it and what is its functional worth? |

| Procedure | Statement |
|---|---|
| | May I remind you that cost of whole car is about Rs.4,00,000/-, and all that the boot carries is a stepney and occasionally some luggage while going to or coming from a Railway Station or Airport. |
| C.B.<br><br>Get agreement on functional worth & put down the figures (Rs.) in the last column.<br><br>Compare the estimated/actual cost with function worth and point out the value opportunity. | |
| | Reverting back to our example of 2-pin plug, let us allocate worth to different functions on these lines i.e. by comparison or arbitrary allocation. |
| Get consensus and write down the 'worths' in the last row (below) costs – in the table – VA/08/VI.<br><br>Point out the differences in actual cost and functional worth and i.e. value opportunities and rank them as (1), (2), (3), etc. | |
| **JOBS FOR PRACTICE** | |
| | Now, before closing the session, I would; like to know whether you have succeeded in selecting appropriate jobs for our workshop.<br><br>Would you please briefly outline the nature of the project? |
| Ensure that everybody has selected a project for class work. Collect details of at least one project from each group. | |
| | We will need these jobs for practice in our next session. If any details, drawing or data are missing, please collect the same before hand. |
| Clean the Chalk Board<br>(except column 1). | |

| Procedure | Statement |
|---|---|
| Close session with appropriate statement:<br><br>  – indicating the programme for next session.<br><br>  – expressing thanks for today's contribution. | |
| | Tomorrow, we enter into a very Creative Phase of our programme. |
| Distribute VM-03/VI | The techniques practiced in that session will be useful not only in VA/VE but in any other area for problem solving. |
| | Thank you very much for your contribution to the deliberation. |

# NOTES FOR SESSION – VI

Ref. to FUNCTION ANALYSIS SYSTEM TECHNIC (FAST) is to be included for advance course/repeat course.

Definition of function using Verb + Noun: As far as possible, verb to be "active" (not like "facilitate," but like transmit, hold, etc.)

Similarly noun should be "measurable".

Active Verb + Measurable noun in definition of function indicate WORK FUNCTION, otherwise Sell function.

Test of primary or basic function & secondary Function.

Training Film No.1:     FUNCTION, COST, WORTH.

# 8

# VALUE ENGINEERING WOREKSHOP

## SESSION - VII
## — SPECULATION PHASE: PART-I
## (Outline of Creativity)

# VALUE ENGINEERING WORKSHOP

## Session-VII:- SPECULATION PHASE : PART -I

## (Outline of Creativity)

| Procedure | Statement |
|---|---|
| <u>Open session:</u><br><br>No need of reviewing previous session. | |
| <center>**CREATIVITY**</center> ||
| Do not labour to offer a full course on "creativity".<br><br><u>OBJECTIVES:</u><br><br>– to make the clear meaning of creativity<br>– to bring out factors that fosters & hinder creative output<br>– to illustrate how to be deliberatively creative.<br><br><br><br>Put a tick mark on item 3, <u>"SPECULATION"</u> in the first column of C.B. and write "3: SPECULATION" in 2nd column as heading.<br><br><br><br>C.B.(last column)<br><u>"CREATIVITY"</u><br><br>Lecture method | In this session we enter into a very interesting phase called "speculation or creative phase".<br><br><br><br>**What sets man apart from all other forms of life is power of imagination or creativity.**<br><br><br><br>Creativity is one of the elements of complex compounds of many interwoven abilities called intelligence, which itself is something very marvelous and defies precise definition. |

| Procedure | Statement |
|---|---|
| | Let me hasten to add here that a creative person is not necessarily the one who gains the highest marks in intelligence tests or who can talk expertly and fluently on any subject under the sky. |
| | Creativity simply means doing our jobs, solving our problem in a better way, a new way (which need not be new to others). |
| | Alex F Osborn defines **Creativity as "ability to visualise, to foresee, and to generate ideas".** |
| Under the heading "Creativity" write:<br><br>– To visualise<br>– To foresee<br>– To generate ideas | |
| | According to Abraham Maslow, "To be creative is to be able to confront novelty and to improvise while enjoying change." |
| | Individual creativity may be defined as capacity to suggest a great quantity of novel and worthwhile ideas based on different principles or approaches; |
| | OR |
| | Adeptness at doing a thing in several ways or finding many alternatives/solutions to a given problem. |
| Factors on Creativity: | Who is creative person? Is it an Inborn trait? Does it depend on age, education or sex? |
| (Age, race, sex, education, and effort.) | |
| (Lecture method: contd.) | |

| Procedure | Statement |
|---|---|
| | Well, a new wave of research in Psychology is shattering the old popular notion that creativity is something inborn or inspirational. |
| | Creative talent is 'normally' distributed. It is ageless and does not discriminate between young and old, nor there is any direct relationship between creativity and education or creativity and sex. |
| | In some areas like arranging parties or managing house-hold affairs, the idea fluency of women is higher than that of male. In scientific and engineering fields, however males score more than females. |
| | Education is also not a vital factor. |
| | Higher educated persons may be sterile creatively and uneducated persons may come up with brilliant ideas. |
| | Most of the famous creative achievements by the inventors and writers were accomplished by them when they were in their 70's or 80's. |
| | If creative talent shows signs of dimming down with advancing age, it is not because of age but <u>lack of efforts.</u> |
| | Factor of effort plays most important role in creativity and with advancing age, the effort dies first and creativity appears to dwindle. |
| | Everybody is a creative minded person. |
| | Imagination is our human birthright. |
| | Of course, it may be dormant, waiting for you to arouse it from deep slumber. |
| | Creativity exists where there is DESIRE and WILL to create. |
| | If it is so, why everybody is not 'top' in generating new ideas? |
| | Let us see why our thinking potentials remain untapped or dormant. |

| Procedure | Statement |
|---|---|
| | What affects our creative output and how can we be deliberately creative when we want to be. |
| **<u>EXERCISE ON CREATIVITY</u>** | |
| No definite set rules can be prescribed for conducting sessions on Creativity. The structure and contents can be varied, and the conclusions arrived at, in a countless, creative ways.<br><br>In fact, the methodology should reflect your own resourcefulness and high degree of creativity.<br><br>The number and type of examples and exercises to be given in this session will depend upon availability of time and fields of experience of the participants.<br><br>Please go through the exercises, explanation and hints, before hand.<br><br>Copy down the gist on a card.<br><br>Present the problems and draw out the objectives in an informal, natural way.<br><br>Conduct the session by using a card, preferably closing the manual. However, ensure that the exercises bring out the objectives, namely, to develop:<br><br>– greater 'curiosity':<br><br>– attitude of 'self-confidence':<br><br>– strong 'motivation':<br><br>– 'open mindedness' to the ideas of others;<br><br>– 'greater sensitivity' to the problems; | |

| Procedure | Statement |
|---|---|
| – divergent/convergent thinking.<br><br>– to improve potential for developing useful ideas.<br><br>Utilise the solutions and the interaction which may be expressed eloquently, humorously, or hot tingly in identifying and high-lighting the mental or perception blocks, cultural blocks, emotional blocks, and habitual blocks. | |

| **WARM-UP EXERCISES** | |
|---|---|
| – Here are a few sample exercises.<br><br>– Design your own exercises on similar lines, varying the themes from batch to batch.<br><br>These examples bring out "Idea Fluency" and show how we underestimate our own capabilities.<br><br><br><br>Watch for the poor response.<br><br>Now, ask them to jot down as many words as they can.<br><br>Allow 7/8 minutes and ask the scores. Many may score above 20. | <br><br><br><br>How many words can you recollect which start with 'R'?  Please raise your hands if the answer is "above 20".<br><br><br><br>This only shows that we under-estimate our capabilities.<br><br>i)  Next, think of all the objects you can that start with letter (a) 'B' (b) 'C' (C) 'S' (any one). |

| Procedure | Statement |
|---|---|
|  | ii) Think of all the objects that are (a) blue or (b) green (any one). |
|  | iii) Think of all the uses of (a) paper, (b) old tooth brush, or (c) paper Clip (any one). |
| Allow 10 minutes for each example (one or two examples are enough) |  |
| **MORE EXERCISES** |  |
| Exercises to illustrate different abilities, that is:<br><br>– ability to see unfamiliar things in a familiar thing;<br><br>– ability to view the familiar in a unique way;<br><br>– ability to see multiples of things in single object;<br><br>– ability to fantasize;<br><br>– idea fluency, originality, etc.<br><br>While the participants are answering the questions or suggesting alternatives, watch for and make a mental note of, critical attitude of some participants for the suggestion of the other participant(s) either in the form verbal attack or gesture or sounds disapproving that idea as something impracticable, ridiculous, or childish. Tactfully prevent such criticism.<br><br>Ex.1 & 2 pertain to imaginative interpretation of known facts. |  |

| Procedure | Statement |
|---|---|
| | 1. How many words can you discover in 'AFFECTIVE'? |
| <u>C.B.</u>: AFFECTIVE<br>Allow 5 to 10 minutes. | |
| | 2. Construct 4-word sentences from the initial letters:<br><br>W—H—A—T—? |
| <u>C.B.</u>: W --H -- A --T | |
| | For ex. Who Has Arrived Today?<br>Whale Hit At Torpedo. |
| Ex. 3 & 4 illustrate the ability to fantasize. They are situation type creatively phrased problems which evoke multiplicity of answers. | |
| | 3. What if you were (a) a cloud,<br>(b) a flower,<br>or<br>(c) a thunder bolt?<br>(any one).<br>OR |
| | 4. If all the water were removed from the earth? |
| Response is likely to be limited to 3 or 4 ideas. | In a group of sixth graders similar problem evoked a dramatic response. |
| | A child was pretending as if grasping for the breath. |
| | A girl was hopping like a frog (imagining that water in her pond had dried). |
| | Another boy was signalling all his relatives by signs to come to its "death" bed. |

| Procedure | Statement |
|---|---|
| | Yet another boy was closing his nose with hand kerchief (indicating stink emanating from decaying matter), and so on.

Comparison between the responses brings out the fact that children symbolise freshness, and aliveness, sharing excitement in exercising their imagination.

These attributes are essential for stimulating creativity. But unfortunately with every passing year we allow these attributes to diminish in geometrical proportions, resulting in poor crop of ideas for even a very fertile problem like this. |
| Ex.5 creates awareness for creative 'looking' & flexes mental muscles. | 5. Arrange 4 match sticks of square cross section and <u>unequal</u> lengths to form a <u>square.</u> (no bending, no cutting, and no side. protruding beyond its adjacent side). |
| Allow 5 minutes. The chance of getting solution from the participants is 1 in 1000.

Disclose the solution.

<u>Solution to Ex.5:</u>

In one solution, the match sticks are arranged as shown below:

| |

| Procedure | Statement |
|---|---|
| In 2nd solution, the sticks are bundled and viewed from the end: In fact, the problem can be made more challenging if a CUBE is to be formed by using the same 4 match sticks. Again, the solution lies in the 1st figure given on p.129. The end of each match stick is a square and 4 such squares stand vertically resulting in a figure of hollow cube. Ex.6 demonstrates how our thinking potential is dormant and untapped, and also the habit of thinking in 2 dimensions. Call two members to your side. Supply 7 round coins and ask them to arrange as per 'statement'. The rest of the participants to try on paper. Ex.6: It will be a miracle if any participant succeeds in solving the problem. | In searching for a solution in routine way, we tend to look for something big, something usual, and miss the small but vital aspect. 6. Arrange 7 coins in 2 lines so that there are 4 coins in each line. Next, move and rearrange one of the coins in such a way that one line has 5 coins and still the 2nd line has 4 coins. |

| Procedure | Statement |
|---|---|
| Demonstrate how easily the task could be fulfilled and point out that if, and only if, they were thinking outside the 2 dimensions in which 7 coins lay and inside 3 dimensions could they succeed. (Block = Habit).<br><br>Disclose the solutions:<br><br><br><br>Initially you may start With 5 coins arranged in 2 rows in the shape of letter 'L' with 3 coins in each row. If the constraint imposed by the shape, 'L' is removed, the solutions will result in not one but many configurations like these:<br><br><br><br> | Well, all you have to do is to pick up one of the end coins and place it on the top of the coin common to both the lines. |

| Procedure | Statement |
|---|---|
| Ex. 7 & 8 show our ability to see multiple of things in a simple object. It also brings out "flexibility" and "originality". | 7. What all things and ideas come to your mind by looking at:<br><br>(a) Δ      (b) 'X' (any one) |
| Draw either a triangle or an 'X' on the chalk board (any One).<br><br>Ex.7:   (Ideas from other groups).<br><br>A consolidated list of ideas that emerged from the earlier 3 sessions in respect of 'x' is given below:<br><br>1. Month of October.<br>2. UK PM's residence at Downing street.<br>3. Accident (bandage).<br>4. Roman letter 'x' (ten).<br>5. Prohibited (smoking, blowing horn, etc.)<br>6. Cross roads.<br>7. Multiplication.<br>8. Wrong answer.<br>9. Tally account.<br>10. Shares in Company (stock exchange).<br>11. Gear shifter in 2-wheelers.<br>12. Appointment with the Doctor (Red Cross symbol).<br>13. Pair of scissors.<br>14. Christmas festival.<br>15. Marketing attendance.<br>16. Reading of scriptures (book stand).<br>17. Jail (cross section of a chain).<br>18. Napkin holder.<br>19. X-ray.<br><br>Mark and point out a light wave of laughter or direct comments on ideas like "red cross," scissors, jail/chain, etc. | |

| Procedure | Statement |
|---|---|
| The commentator(s) may also advance argument as to how red cross symbol is different from symbol 'x' drawn on the board.<br><br>Ask the participants to mark such comments and request not to evaluate others' ideas at this stage. Remind them that evaluation phase follows after this session.<br><br>On conclusion of idea producing spell of 10 to 15 minutes, brings to the notice the wordings of the problem i.e. "**what comes to your mind** by looking at 'x'? and that the problem was **not as to "what looks like an 'x'**"?<br><br>Also utilize the comments and remarks of participants in brainstorming sessions (that follow later) to illustrate as to how pre-judgement and criticism acts as barrier to the flow of ideas.<br><br>C,B.<br><br>Draw two concentric squares and shade the inner one.<br><br><br><br>This exercise is on "development of alternatives". But the solutions will also expose the road blocks like "lack-of-concentration," "will-to-confirm," "fear-of-failure" (i.e. anxiety to be the first in answering the question), etc. | 8. Divide the un-shaded area into 4 equal parts of the same shape. |

| Procedure | Statement |
|---|---|
| Invariably the participants will join the mid points of the sides or the nearest corners of two squares by <u>straight lines</u> and stop after getting two or three solutions. | |
| You may be wondering as to what is so great about this problem and what is wrong with the above mentioned solutions. By the way, do you know how many solutions are there to this problem? | |
| Have you guessed, guessed, and failed to imagine more than a dozen solutions? | |
| How many POINTS are in a line ? | |
| Well, all that you have to do is to count the number of 'points' between the corner and the mid point or between the corners of any side of the figure and join corresponding points of the 2 similar figures. By this time you are convinced that there are <u>infinite</u> number of <u>solutions</u> to the problem (by using the corresponding points). | |
| If you stretch your imagination further and ask yourself as to why to join only the corresponding nearest points, and why not point A with B and so on, you will find <u>another series of infinite solution.</u> | |
| That is not all to the solutions. Again ask yourself "why join the point by straight lines?" why not curved lines? | |
| Zig-zag lines? Bent lines? and so on. | |

| Procedure | Statement |
|---|---|
| | This example illustrate the awful truth that our creative potentials is latent and untapped.<br><br>When we attach a problem as soon as 'a' solution comes to our mind, we feel satisfied and stop exercising our imagination in search of more and newer ways of doing a thing.<br><br>(Blocks: Mental laziness, anxiety to be "the first" with solution, thinking in terms of old solutions). |
| Ex.9 & 10 are directed towards generating awareness for improvement, improvisation, innovations. They should trigger an inner urge and strong will to identity problems for creative attach (problem sensitivity). | |
| | 9. What improvements would you like to make in:<br>a) rubber slipper,<br>b) in a gas stove, or<br>c) shirt button? (any one). |
| Record ideas on Chalk Board<br><br>You may give lead by stating that you would like:<br>(1)  to drive a bicycle by wind power:<br>OR<br>(2)  to make water repellent coating on wind shields of cars so that no whiper is needed;<br>OR<br>(3)  to invent a chalk board on which matter can be written by finger or wooden stick (No chalk, No dust). | |

| Procedure | Statement |
|---|---|
| Get at least five such "felt-need" problems and record on C,B.<br><br>Point out to "ideas" recorded in response to Ex.9 & 10. | 10. What inventions would you like to have to solve every day problem?<br><br>Your contribution should make one thing clear:<br><br>Knowing what we are looking for helps us to recognise it when you look for it enquiringly.<br><br>Now, a problem for creative attach can be recognised in two ways:<br><br>No. one:   by redefining the problem consciously, and<br><br>No. two:   first by stating the problem as broadly as you can, and then by breaking down this problem into a no. of specific problems. |
| Exercises on:<br>– Problem redefinition/Restructuring;<br>– Broadening a problem;<br>– Breaking down the problem into sub-problems. | Let us examine the advantages of specifying our problems consciously of using synonyms is multi-expression of the problems and of spitting a problem into specific sub-problems.<br>11. Your friend has passed a coveted degree exams, standing first in the State.<br><br>How would you begin a congratulatory letter to him? |

| Procedure | Statement |
|---|---|
| <u>Allow a few minutes:</u><br><br>Invariably, the response will be something like this:<br><br>   – Hearty congrats …..<br><br>   – Immensely pleased to know …..<br><br>   – Delighted to know …..<br><br>   – It was a pleasant surprise to know…..<br>     etc. etc.<br><br><br>Without comments on the participant's contribution, continue further. | |
| | Now, I will rephrase the same problem.<br><br>"In how many ways can you convey your congratulations to your friend?" |
| <u>Allow 5 or 6 minutes:</u><br><br><br>Expected response would now run something on the following lines:<br><br>– Meet him personally;<br><br>– Send a standard printed card;<br><br>– Talk to him on telephone;<br><br>– Send a letter;<br><br>– Convey your message through somebody visiting his place;<br><br>– SMS or email and so on. | |
| | The first statement of problem limited the choice of communication to only one method viz. 'writing'.<br><br>By re-phrasing the same problem, we could get a multiplicity of communication ideas. |

| Procedure | Statement |
|---|---|
| | 12. Another examples: |
| | A mother of 7 year old boy has been complaining that her son does not take milk. In fact the dislike the very sight of a milk glass. |
| | Could you please help this lady by suggesting as to how to make the boy drink milk? |
| Allow few minutes:<br><br>Anticipated suggestions would be as follows :<br><br>– Serve porridge (kheer) of …..<br><br>– Mix egg or chocolate or 'Viva' or 'Horlicks' to milk<br><br>– Coax by offering chocolates or new dress every month.<br><br>– Mix syrup in milk/shake<br><br>– Hold out threat of punishment like…..<br><br>"No permission to play with friends unless ….."<br><br>"No movie unless….."<br><br>"No picnic ……," etc. | |
| | Let me tell you who solved this problem and how. |
| | In a "coffee party" when the mother was narrating her woes to her friends, the maid servant, who was fed-up of listening this woe repeatedly, asked suggestively; |
| | "Why do you want (the boy) to drink milk?" |

| Procedure | Statement |
|---|---|
| | Provoked by the unexpected question from the most un expected quarters, the ladies took it as a challenge to their education, status, etc. and thought aloud.<br><br>"Yes, why should milk be trust down the throat of the boy? why indeed?"<br><br>They redefined the problem by answering the maid's question like this:<br><br>"Milk is required to provide body building proteins and nutrients."<br><br>Correctly stated, the problem then became – what else can supply these nutrients?<br><br>Well, there was a long list of useful suggestions for the mother and the child to be happy thereafter.<br><br>The above two examples should make it clear that problem is not "how to make a child drink milk" or "how to write a congratulatory letter" but how to specify our problems consciously.<br><br>A problem well stated is half solved. |
| PAUSE: | Coming to the 2nd aspects, namely, splitting of all inclusive problems into specific sub-problems, let us take one or two examples.<br><br>13. A manufacturing organisation is plagued with low productivity.<br><br>A broad statement of the problem as to "how to improve overall productivity" in the organisation will lead us no-where. |

| Procedure | Statement |
|---|---|
| | Instead, if we break it up into sub-problems, it can help us in focussing our attention on specific areas. |
| | Thus we may ask ourselves: |
| | − In how many ways can we reduce inventory of : <br> a) Raw-materials; <br> b) consumables; <br> c) finished goods? <br> − In what way can we increase utilization of: <br> a) machinery & plant; <br> b) materials; <br> c) labour? <br> − In how many ways can we cut expenses on energy? <br> − In what way can we increase turnover or attract more customers? |
| | And so on. |
| | In fact, each of the sub-problems can be further sub-divided into several specific questions for creative attack. |
| Do we need to increase or reduce staff? may invoke a mere "yes"/"no"/"perhaps" answer. Such questions are a matter of judgement and not for a creative attack. | |
| | Let us take one more example of problem stated in broad term: |

| Procedure | Statement |
|---|---|
|  | 14. The problem is "How to reduce fuel consumption in a car?" <br><br> What are possible sub-problems? |
| Make a mental note or briefly record answer on C.B. (last column). |  |
| NO discussions, no comments, and no elaboration required. |  |
| Announce coffee or tea break. <br> Clean C.B. (except col.1&2) |  |
| Distribute Hand Outs VM-04 to VM-08 (5 Nos.) <br> Pp. 248 to 258 |  |

# 9

# VALUE ENGINEERING WORKSHOP

SESSION - VIII

— SPECULATION PHASE: PART-II

(Brainstorm Session)

# VALUE ENGINEERING WORKSHOP

## SESSION-VII: SPECULATION PHASE : PART- II

## (Brainstorm Session)

| Procedure | Statement |
|---|---|
| Before conducting a Brain – storming, be advised to be reasonably familiar with the various techniques of Creativity Development, such as:<br><br>– Morphological Analysis;<br>– Forced relations technique:<br>– Synectics technique;<br>– Work simplification, etc.<br><br>(See Pp. 148–150 appended at the end of this session notes).<br><br>OPEN SESSION.<br><br>DO NOT REVIEW PREVIOUS SESSION. | |
| **CREATIVITY TECHNIQUES** | |
| Lecture Method:<br><br>*(    )*        This portion is optional. | *(There are a number of procedures for stimulating the flow of ideas and improving the output of novel and worthwhile suggestions, such as:<br>– Morphological Analysis;<br>– Forced Relation Technique;<br>– Synectics; etc.<br>– <u>Morphological Techniques:</u> which are also known as analytical creativity techniques. They are based on a logical analysis of the problem and its various elements. |

| Procedure | Statement |
|---|---|
| | – <u>Forced Relation Technique:</u> in which forced relations are established between two or more elements, ideas or subjects which are seemingly un-related. |
| | For ex.    "ship" and "space" <br>        = "spaceship" |
| | – <u>Synectics:</u> is another method of creative problem solving in which a client who has a problem, also participates; and it involves personal, direct, and symbolic analogies. |
| | – <u>Group Brain Storming:</u> One of the simplest and most effective technique suited for Value Engineering workshop is Group Brainstorming. |
| | The approach involves statement of problem, first by the group leader or facilitator, then by the group, and lastly, selecting of the best statement. |
| | The emphasis is on generation of a good number of ideas in an atmosphere free from constraints, negative responses, critical judgements, and boundary conditions imposed by traditional thought process. |
| | You may recall the basic rules of brainstorming that were practised by the group led by Mr. Bright. <br> They were: |
| | – Suspended judgement, i.e. no criticism, no evaluation of any idea. <br> – Wilder and original ideas are welcome. <br> – Quantity is encouraged. <br> – Cross fertilisation freely allowed)* |

| Procedure | Statement |
|---|---|
| **BRAIN STORMING RULES.** | |
| C,B, Column-2 under "Speculation," write:<br><br>BRAINSTORMING RULES.<br><br>1. Defer judgement;<br>2. Welcome free wheeling;<br>3. Aim for Quantity;<br>4. Combine & improve ideas. | Let us adhere to these principles strictly.<br><br>Let the ideas flow freely, without any hindrance, and judgement on their utility or futility. Please remember the time for evaluation comes later.<br><br>Secondly, wilder and even seemingly impractical ideas are welcome<br><br>Let us also remember that quantity helps bread quality. The greater the number of ideas flowing, the more chance of striking original ideas.<br><br>The combination of ideas yield a new crop of ideas.<br><br>We will now proceed to put these principles in practice. |

## CASE STUDIES ON 'SPECULATION'

| Case Study No.2; | |
|---|---|
| | This is a live case-study conducted by a group of 5 persons in a VE/VM workshop conducted by Indian Institute of Management, Bangalore. |
| Distribute Pp 1, 2 & 3 and Annexure-I of the Handout-VA-10/VIII in this session and the rest, after 'evaluation'. | |

| Procedure | Statement |
|---|---|
| Allow 10 to 15 minutes. Clear doubts. | |
| | Based on the available data on the first two phases, viz. Selection & Information, let us now search for alternatives to achieve the given functions. For this purpose, as before, we will work in teams. |
| Retain the composition of teams as formed in earlier sessions. | |
| Explain sitting arrangements and time allowed (about 20 minutes). | |
| | The team leaders will ensure that each & every idea is recorded, and that everyone, including himself, strives hard to suggest several alternatives. |
| | At the end of the allowed time, the group leaders will prepare a list of ideas and handover to me. |
| Collect the lists of ideas and list them in the 3rd column of blackboard, eliminating duplicates. | |
| Refer to Annexure-II of the handout VA-10, and see whether all ideas are covered by the list on C.B. | |
| If not. and if you have your own ideas, add them to the list after taking formal approval of the participants. | |

| Procedure | Statement |
|---|---|
| – Point to list on C.B.<br>– List additional ideas. | (If you permit me, I would like to be one of you as a participants and contribute my ideas to this list.)<br><br>These may not be the final lists.<br><br>You may get new insights during 'incubation' period.<br><br>You may think of combining two or three ideas; or improving upon some of them.<br><br>Please, jot down such "flashes" as and when they occur. We will finalise the lists tomorrow. |

| Procedure | Statement |
|---|---|
| **MEMBERS' JOBS** | |
| MEMBERS' Problems:<br><br>(See notes on Pp. 151 & 152. of this session notes).<br><br>Take at least two problems for Brainstorming from the projects collected at the end of Session-VI.<br><br>Brainstorm the first problem, involving all the participants as a single panel.<br><br>  – Sit down & record ideas on a sheet of paper.<br>  – At the end of allotted time, read out the listed ideas.<br>  – For the 2nd problem, divide the members into teams and follow the same procedure as in case studies 1 and 2. However, the ideas need not be recorded on the C.B.<br><br>Teams re-assemble.<br><br>Close Session.<br><br>Retain Board work.<br><br>FILM SHOW ON CREATIVITY: | |

# NOTES FOR SESSION-VIII

## I. DEVELOPMENT OF CREATIVITY:

The starting point of any improvement, innovation, or invention is generation of worthwhile idea. But generation of an idea is in itself a complex psychological process and there is no generally accepted theory or the model available to suggest the possible way of idea generation. However, the trainer should make himself familiar with some of the famous models like:

– Perception model involving both sensory and extra sensory perceptions.

– Fusion model in which information is the key concept and is based on the assumption that if a technical opportunity is recognised, innovative ideas spring up to fulfil the perceived need.

The various stages of development of creativity include:

1. Administering creative test which is a tool for measurement of creative potentials of individuals i.e. capacity to suggest a number of worthwhile ideas based on different principles and approaches.

2. Measurement of creative climate in any organisation as perceived by the participants.

3. Application of techniques to be deliberately creative. Such technique include:

   (a) Morphological Technique: It is an analytical tool of ensuring that all possible solutions to a problem are taken into account through a systematic breakdown of a problem into its sub-problems, which can then be attacked independently.

   (b) Brainstorming: It is method of creative thinking based on free association and deferred judgement. (See text in manual.)

   (c) Forced Relation Technique: In this technique forced relationships are established between two or more ideas or objects or elements which are seemingly unrelated according to habitual thinking patterns.

      Most commonly used approaches are:

      i) Catalogue technique;

      ii) Listing technique;

      iii) Focus –object technique.

      The element chosen are considered in all possible combinations as a basis for free association from which novel and practical ideas will emerge.

(d)    Synectics: It is a method of problem solving where one attempts to simulate the thinking process when the individuals are most creative. The participants include a client who has an unsolved problem and a group of member trained in the method. The procedure involves statement of problem as given by the client, as by the group, stimulation of divergent thinking by putting evocative question which leads to personal analogy, direct analogy, or symbolic analogy in obtaining a solution.

(f)    Work Simplification: Work simplification is a special application of a Method Study where the employees are trained to analyse and improve the work they perform, in a systematic manner.

(g)    Quality Circle: Its essence is team approach at grass root level where the working level people are encouraged to identify quality problems, suggest improvements and implement the ideas. Quality being synonymous with Value, Quality Circles aim at improving Value of the operation, product or service – in fact, the ultimate aim is to improve 'quality of life' through numerous suggestions.

# NOTES FOR SESSION-VIII (CONTD.)

## II. BRAINSTORMING OF MEMBERS' PROBLEMS:

1.  After collecting the 'projects' from the participants at the end of Session-VI, and before conducting the Brainstorms, you as a panel leader or facilitator should ensure:

    (a) that between the sessions you have visited and studied it in working situation;

    (b) that you have all the necessary 'information' on hand;

    (c) that the 'Information' is analysed and arranged systematically (up to the end of information phase);

    (d) that the copies of the 'information' are available for distribution, before brainstorm.

2.  For brainstorming the members' problems, you should develop in advance your own list of 'ideas'. so that if the flow of ideas slows down, you can prime the flow by interpolating ideas of your own.

3.  You should also be well prepared to give hints in the form of 'spur' questions like:

    – <u>Why</u> is it necessary?

    – <u>What</u> else can achieve this function?

    – <u>Where</u> should it be done?

    – <u>Who</u> should do it?

    – <u>How</u> should it be done?

    You should kindle the fire of imagination by strokes such as:

    – What about …?

    – What if ……..?

    – What else? and what else ……..?

    - Why not eliminate ……?

    – Why not combine …………..?

    – Bigger instead of smaller ……?

    – Smallest instead of bigger ……?

    – What different materials to achieve this result …….?

    – Why not reverse………………………………………….?

    – Why not rotate . . . . . . . . . . . . . . . . . . . . . . . . . . . . . . ?

    – Why not shift location ………………………………….?

    – Is there something similar………………………… ?

    There is an endless list of such questions, and you should prepare your own list of specific questions that help in activating imagination in the right direction.

    You may copy the 'spur' questions on a card and use the card in the brainstorm sessions.

# 10

# VALUE ENGINEERING WORKSHOP

## SESSION - IX

## — EVALUATION PHASE: PART-I

# VALUE ENGINEERING WORKSHOP

## Session-IX: - EVALUATION PHASE: PART - I

| Procedure | Statement |
|---|---|
| <u>OPEN SESSION</u><br><br><br><br><br><br><br><br><br><br><br>Point to C.B. 3<sup>rd</sup> & 4<sup>th</sup> columns. | In the previous session we have seen how to be deliberately creative, and how to generate multiplicity of ideas.<br><br>The objective of brainstorming session was to <u>get ideas,</u> and not to solve the problem.<br><br>Two brainstorming sessions yielded these ideas. |
| **SELECTION OF CRITERIA** | |
| Record the new 'ideas' on PAPER REEL CORE under existing ideas on C.B., and also on the lists that are with you. | You might have thought of some more ideas. Will you please spell them out?<br><br>In addition, we have two more lists of ideas on your 'jobs'.<br><br>By your judgement and experience you are likely to reject some of the listed ideas summarily, being too wild or ridiculous.<br><br>You may also be tempted to recommend certain ideas for straight-away implementation.<br><br>Rejection or acceptance of ideas based on judgement and experience will be totally **subjective** and will defeat the very purpose of generating a host of ideas. |

| Procedure | Statement |
|---|---|
| | What is needed at this stage is an unbiased and **objective** treatment in the form of numerical method. |
| | In this session we will practise some important techniques of judging the relative effectiveness of ideas and arriving at the optimum solution. |
| Clean the 3rd & 4th column of C.B. | |
| Put a tick mark on the 4th heading i.e. "EVALUATION," in the 1st column. | |
| In the 2nd column write: | |
| 4: EVALUATION: | |
| Objective: | |
| To practice selection of criteria of evaluation and their ranking in order of importance. | The factors, attributes, or criteria to be considered in the evaluation will, ofcourse, depend upon: |
| | -the complexity of the problem or the function; and |
| | -the degree of refinement sought to be achieved in evaluation. |
| | Accordingly, the process of evaluation will vary from a simple technique of listing good and bad points to the complex method of matrix analysis. |
| | Whatever be the technique, the first step in evaluation is the <u>selection of criteria</u> that must be considered in comparing the effectiveness of different alternatives. |
| <u>C.B.:</u>    <u>Criteria</u>: in 2nd column under | |
| "EVALUATION". | |

| Procedure | Statement |
|---|---|
| | What factors do you take into account before buying a consumer durable, say, the bicycle or pressure cooker. |
| Bring out and record in the 2nd column under "Evaluation" factors like:<br><br>– Quality;<br>– Reliability;<br>– Performance;<br>– Price;<br>– Availability;<br>– Durability;<br>– Serviceability;<br>– Ease of operation;<br>– Safety;<br>– Appearance, etc.<br><br>Statements like "consult Friends," or brand names should be related to some of the attributes listed above. | |
| | What additional attributes should be considered in case of an electrically operated appliances like a table fan, or a washing machine? |
| Record additional factors only on the C.B. (like power consumption, capacity, electrical reliability, etc.) | |
| | And selection of fridge? |
| Just listen and push on. | |
| | So, it is advisable that our first task in evaluation is to prepare a check list of common yardstick with which to measure the comparative merits of ideas. |
| Distribute Handout-VA-12/IX (pen holder). | |

| Procedure | Statement |
|---|---|
| Case Study-4 (Pen holder) | Here is a list of ideas suggested by a group of technician apprentices for a ball-point pen holder. |
| Allow few minutes to study the illustrations. | What are the possible attributes that are appropriate for evaluating these ideas? |
| If necessary use suggestive questions to bring out about 5 or 6 factors amongst the following:<br>– Material;<br>– Number of parts;<br>– Number of operations;<br>– Special tools & skills;<br>– Weight;<br>– Aesthetic;<br>– Stability;<br>– Durability;<br>– Safety;<br>– Design cost. | |
| | ( – How about number of parts?<br>– Does it effect the manufacturing cost?<br>– Assembly time?<br>– Any special tools and skills required to make and/or assemble?<br>– Do you think that string will last long? (durability)<br>– What about safety? (sharp edges)<br>– Will the tumbler be stable at all times? (stability)<br>– Which design needs least amount of material and operations?<br>and so on.) |
| List 5 or 6 characteristics on C.B. (3rd column) and number them as A,B,C,D, ......... | |

| Procedure | Statement |
|---|---|
| | Which of these are very important or less important and in what order?<br><br>Will you please help me in ranking these factors in terms of their relative importance? |
| Get the ranks quickly and record against the criteria A,B,C,D, …….<br><br>The board work now appears some what like this:<br><br>Key | |

Key

| Letter | Factor | Rank |
|---|---|---|
| A | No. of parts | 1 |
| B | Durability | 2 |
| C | Safety | 4 |
| D | Convenience | 3 |
| E | Material cost | 1 |
| F | Design cost | 5 |

## MATRIX TECHNIQUES OF EVALUATION

| Procedure | Statement |
|---|---|
| **MATRIX SYSTEMS:**<br>(Forced Decision Techniques)<br><br>Explain by lecture method.<br><br>Draw the matrix in lower half of 3rd column of C.B. | Let us examine whether this ranking by 'hunch' or judgement, stands the scrutiny of a numerical method. |

| Procedure | Statement |
|---|---|
| | In the matrix systems of determining the 'weightages' is based on the fact that no two factors have exactly the same merit or value i.e. one of the two factors must be more important than the other, of course , to a varying degrees. |
| | The difference in importance may be minor, medium or major, and is denoted respectively by the digits, 1, 2, and 3. |
| C.B. (3rd column, below matrix). Difference in importance: "1. Indicates minor difference" "2. Indicates medium difference " "3. Indicates major difference" | |
| | The factors are compared in pairs and degree of difference is indicated by the key letter of the more important factor followed by digit 1, 2 or 3, depending on the degree of difference. |
| Write faintly A2 in Matrix. | Thus, in comparing A with B, if A is more important than B, and the difference is medium, A2 is entered into the square of matrix which is common to Row A and column B. |
| Write faintly F3 in Matrix | Similarly, if D is less important than F and if the difference is major, F3 is entered into the square which is common to row F and column D, and so on. |
| | May I mention that the number of comparison (x) to be made in a set of 'n' factors taking 2 at a time, is given by the formula: |
| | $$X = \frac{n(n-1)}{2}$$ |

| Procedure | Statement |
|---|---|
| <u>C.B.:</u> $$x = \frac{n(n-1)}{2}$$ $$= \frac{6\,(6\text{-}1)}{2}$$ $$= 15$$ | This formula is used in determining the number of matches (X) to be played between 'n' teams (in hockey or foot-ball league). In our case n = 6. $\therefore$ $$x = \frac{n(n-1)}{2}$$ $$= \frac{6\,(6\text{-}1)}{2}$$ $$= 15$$ Let us now complete this matrix. |
| <u>PAUSE:</u> Rub off A2 and F3 from the matrix. Ask the participants to compare 'A' with 'B', A with C and so on, and to spell out the degrees of importance. Complete the row A. Next, take B and compare it with C, then with D and so on. Continue till all the factors are compared in pairs. Your completed matrix will look like this: | |

| Procedure | Statement |
|---|---|

**Procedure:**

| | A | B | C | D | E | F |
|---|---|---|---|---|---|---|
| A | | A2 | A3 | A3 | B2 | F2 |
| | B | | B3 | B2 | B3 | F3 |
| | | C | | C2 | E2 | C1 |
| | | | D | | D1 | F1 |
| | | | | E | | F2 |
| | | | | | F | |

Explain the method. First, add the 'numbers' following A, then those following B, etc. Prepare a table in 4th column of C.B. and enter the score in 2$^{nd}$ column of table as follows:

### Weightages

| Key | Points | % | Fraction of 1 | Rank |
|---|---|---|---|---|
| (1) | (2) | (3) | (4) | (5) |
| A | 8 | 24 | 0.24 | 2 |
| B | 5 | 15 | 0.15 | 4 |
| C | 3 | 9 | 0.09 | 5 |
| D | 1 | 3 | 0.03 | 6 |
| E | 7 | 22 | 0.22 | 3 |
| F | 9 | 27 | 0.27 | 1 |
| Total: | 33 | 100 | 1 | |

**Statement:**

The weightage points for different factors are then calculated by adding the numbers following the key letters like this:

Let us compare the rankings of factors by the two methods we have followed so far.

| Procedure | Statement |
|---|---|
| Point to two tables on C.B.<br><br>Compare the ranks given by 'hunch' and arrived at by numerical method.<br><br>If, by any remote chance, the two rankings are exactly the same, give credit for accuracy of "Guestimates" made in the 1st method; and emphasize that such a coincidence is rare. If not, mention that the numerical method leads to objective evaluation. | Numerical methods leave no room for ambiguity and lead to objective assessment.<br><br>There are several ways of extending this matrix techniques to evaluation of ideas generated in brainstorming sessions. |
| **GRADING METHODS** ||
| Criteria **Value-T-Chart-I  or Grading Method**.<br><br>OBJECTIVE:<br><br>To practice use of various grading (numerical) methods of evaluation.<br><br>Explain by lecture method. | One of the simplest ways of evaluation is to first list out the factors or characteristics that are relevant to the case under study and number them as A,B,C,D, etc |

| Procedure | Statement |
|---|---|
| | Next, on a scale of 0-10, designate the grades of each factor as: |

<table>
<tr><td></td><td><u>Grades</u></td><td><u>Points</u></td></tr>
<tr><td></td><td>Excellent</td><td>10</td></tr>
<tr><td></td><td>Very good</td><td>8</td></tr>
<tr><td></td><td>Good</td><td>6</td></tr>
<tr><td></td><td>Fair</td><td>4</td></tr>
<tr><td></td><td>Poor</td><td>2</td></tr>
</table>

C.B.:

| Poor | Fair | Good | V.G. | Xlnt |
|---|---|---|---|---|
| 0    2 | 4 | 6 | 8 | 10 |

Next step is to prepare a matrix like this:

Draw the matrix on C.B

(4th column, lower half)

| Alternative | Design effectiveness | | | | | Score | Rank | Decision |
|---|---|---|---|---|---|---|---|---|
| | A | B | C | D | E | | | |
| A-1 | | | | | | | | |
| A-2 | | | | | | | | |
| A-3 | | | | | | | | |
| A-4 | | | | | | | | |

Each alternative is now evaluated visa-vis the design criteria, and the points entered into the matrix

Thus, if alternative A-1 offers excellent quality and reliability, good service ability and safety but poor productivity, the 1st row of our matrix would read like this:

| Procedure | Statement |
|---|---|

| Alternative | Design effectiveness | | | | | Score | Rank | Decision |
|---|---|---|---|---|---|---|---|---|
| | A | B | C | D | E | | | |
| A-1 | 10 | 10 | 6 | 6 | 2 | 34 | | |
| A-2 | | | | | | | | |

In similar manner, compare the efficiency of 2nd alternative in respect of the design criteria A,B,C, etc. and record the points in the matrix against A-2.

Continue till all alternatives are considered.

Complete the matrix;

Calculate scores;

Rank the alternatives; and

Note the decisions.

The matrix reads something like this:

| Alternative | Design effectiveness | | | | | Score | Rank | Decision |
|---|---|---|---|---|---|---|---|---|
| | A | B | C | D | E | | | |
| A-1 | 10 | 10 | 6 | 6 | 2 | 34 | 3 | Reject |
| A-2 | 8 | 8 | 10 | 10 | 8 | 44 | 2 | Hold |
| A-3 | 2 | 6 | 6 | 8 | 8 | 30 | 4 | Reject |
| A-4 | 10 | 10 | 8 | 10 | 8 | 46 | 1 | Accept |

| Procedure | Statement |
|---|---|
| Exercise on 'Pen Holder' (contd.) | Let us apply this method to the alternatives of the pen holder, taking the Fig.1 as standard for comparison and using the earlier listed design characteristics.<br><br>As before, we will work in teams, and each team will take up 5 alternatives. |

| Procedure | Statement |
|---|---|
| | Team-1 will evaluate effectiveness of Alternatives 2 to 6; Team -2 will Take-up 7 to 11; the 3rd team, 12 to 16 and the 4th team, the rest. |
| | You may complete this exercise in about 15 minutes. |
| Allow 15 minutes. | |
| Clean lower half of 4th Column of C.B. | |
| When the teams reassemble, ask the team leaders to give the group decisions, Note the "most desired" alternatives and short-list them on C.B. as follows: | |

| Sl.No | Alternative no. | Description |
|---|---|---|
| 1. | A-1 | |
| 2. | A-2 | |
| 3. | A-3 | |
| 4. | A-4 | |

| Procedure | Statement |
|---|---|
| | We have these 4 short-listed alternatives for final evaluation, and all of you will form a single panel to complete the exercises. |
| All participants as a single group to evaluate the short-listed ideas. | |
| Draw the matrix on the C.B. (lower half of 4th column) | |

| Alternative | Design effectiveness | | | | | Score | Rank | Decision |
|---|---|---|---|---|---|---|---|---|
| | A | B | C | D | E | | | |
| A-1 | | | | | | | | |
| A-2 | | | | | | | | |
| A-3 | | | | | | | | |
| A-4 | | | | | | | | |

| Procedure | Statement |
|---|---|
|  | How do you rate alternative A-1 as regards the quality?<br><br>– Excellent?<br>– Good? |
| Enter the points corresponding to the grade in the appropriate box in the matrix.<br><br>Repeat the questions in respect of other criteria as applicable to alternative-I, and complete the first row.<br><br>Next, take the other 3 alternatives, one at a time, and subject them to similar enquiry.<br><br>Complete the matrix.<br><br>Ask for decisions, record them in the matrix and read them aloud.<br><br>Clean the lower halves of 3<sup>rd</sup> & 4<sup>th</sup> columns of C.B. |  |
| CRITERIA VALUE-T-CHART – II<br>By Lecture Method:<br><br><br><br>Refer to the score board in the 4<sup>th</sup> column, top half of C.B.<br><br>Read out the first 2 or 3 weightages in fraction.<br><br>In 3<sup>rd</sup> column, lower half, a new matrix is drawn like this: | In an improved version of grading method which is also known as point-score method, the relative weightages of the criteria are first converted into percentages or fraction of '1' like this: |

| Procedure | Statement |
|---|---|
| Prepare a new matrix as follows: | |

| Alternative | Weightage: | A 0.24 | B 0.15 | C 0.09 | D 0.03 | E 0.21 | F 0.28 |
|---|---|---|---|---|---|---|---|
| A1 | | -- | -- | -- | -- | -- | -- |
| A2 | | -- | -- | -- | -- | -- | -- |
| A3 | | -- | -- | -- | -- | -- | -- |
| A4 | | -- | -- | -- | -- | -- | -- |

**Statement:**

Next, allocate score varying between 70 & 90 to each criterion, 70 being the least acceptable design effectiveness and 90, the maximum expected.

On this short scale of 70-90, how do you rate alternative A1 in respect of criterion A (quality)?

**Procedure:**

Record the rated value against A1 (above dotted line) and below A.

**Statement:** What about 'B'?

**Procedure:**

Continue till all criteria are covered for alternative 'A1'.

Repeat the procedure for other alternatives (A2, A3, A4 …).

The matrix now reads something like this:

| Alternative | A 0.24 | B 0.15 | C 0.09 | D 0.03 | E 0.21 | F 0.28 | Score | Decision |
|---|---|---|---|---|---|---|---|---|
| A1 | 90 | 90 | 70 | 70 | 80 | 70 | | |
| A2 | 90 | 80 | 80 | 75 | 90 | 75 | | |
| A3 | 70 | 70 | 80 | 90 | 70 | 85 | | |
| A4 | 70 | 80 | 80 | 75 | 70 | 80 | | |

| Procedure | Statement |
|---|---|
| | Next, the rated values are multiplied by weightage fractions and the scores are written below the dotted line in respective columns |
| | Thus, the product of 0.24 and 90, rounded off to the nearest digit is 22 and is written against A1 in the 1st column. |
| Enter the score 22 in the matrix against A1 below 90. | |
| Repeat the other criteria and fill up the 1st line (against A1). Add the 'products' and Record under 'score'. | |
| | The other alternatives are similarly evaluated. |
| Ask the participants to carry-out the exercise in respect of A2, A3, and A4. | |
| Get the 'products' and complete the matrix. | |
| Read out the 'scores' and ask for 'decision'. | |
| Record decision in last col. | |
| | That completes our Criteria Value-T-Chart. |

The final board work looks like this:

| Alternative | A 0.24 | B 0.15 | C 0.09 | D 0.03 | E 0.21 | F 0.28 | Score | Decision |
|---|---|---|---|---|---|---|---|---|
| A1 | 90 | 90 | 70 | 70 | 80 | 70 | 81 | 2 |
| | 22 | 14 | 6 | 2 | 17 | 20 | | |
| A2 | 90 | 80 | 80 | 75 | 90 | 75 | 88 | 1 |
| | 22 | 12 | 7 | 2 | 19 | 21 | | |
| A3 | 75 | 70 | 80 | 90 | 70 | 85 | 77 | 3 |
| | 18 | 10 | 7 | 3 | 15 | 24 | | |
| A4 | 70 | 80 | 80 | 75 | 70 | 80 | 75 | 4 |
| | 17 | 12 | 7 | 2 | 15 | 22 | | |

| Procedure | Statement |
|---|---|
| Compare the decisions arrived at by the two methods. (They will and should, be the same). | |
| Announce broad outline of programmes for the evening as well as for the next session. | |
| EVENING (HOME WORK) | |
| Evaluation of ideas generated in Brainstorming session: | |
| – 1st & 2nd teams to take up "Core of Paper Reel". <br> – 3rd & 4th teams to take up "Pen Holder". | |
| Advise (1) to first short-list the alternatives (not exceeding 4 for each function), by using simple weighing technique and then (2) to use criteria value-T-Chart for final evaluation. | |
| Programme for the next session: | |
| Two more members' problems to be brainstormed, giving a total of 4 idea lists on members' problems. | |
| Each team will take up one list of alternatives and evaluate their effectiveness. | |
| Close Session. | |
| Clean C.B. (except Col.1) | |

# 11

# VALUE ENGINEERING WORKSHOP

## SESSION - X

### — EVALUATION PHASE: PART-II

# VALUE ENGINEERING WORKSHOP

## Session-x:- EVALUATION PHASE : PART - II

| Procedure | Statement |
|---|---|
| <u>OPEN SESSION:</u><br><br>Announce Session's programme. | In this session, we will first take up 2 or 3 of your problems for brainstorming, and then go ahead with exercises on evaluation, as announced at the end of previous session. |
| **BRAINSTORM MEMBERS' PROBLEMS** | |
| - Two problems from two groups were brainstormed in earlier session.<br><br>- Now, take two more problems, one each from the 3<sup>rd</sup> & 4<sup>th</sup> groups.<br><br>- Call the 1<sup>st</sup> 'problem owner' to the head of the table. (One more member from his team may also join him to assist in recording ideas, etc.)<br><br>- Request him to present the Problem up to "information phase". Ensure that function(s) is (are) precisely spelt out. Encourage him to conduct the brainstorming session.<br><br>- Prompt him to put appropriate 'spur questions' to the 'panel' to evoke spontaneous response, suggesting alternative ways of achieving the defined function(s).<br><br>- At the end of brainstorming session, ask the 'problem owner' to read out the suggested alternatives. | |

| Procedure | Statement |
|---|---|
| – Thank the 'problem owner' and his partner and return them to their seats with a word of encouragement (for the excellent job that they have done). | |
| REPEAT FOR THE "4ᵀᴴ PROBLEM": | |
| | You have a list of ideas to achieve the desired function(s). |
| | You may add, combine and/or improve upon, these ideas later on. |
| | If need be, you can short-list them by using simple weighing technique, before using any grading method of evaluation. |
| **EVALUATION OF ALTERNATIVES** | |
| (Members' case studies) | |
| Team work: | |
| | We have 4 lists of ideas, two from the previous session and two from this session, for evaluation. |
| | As in the previous sessions, we will work in groups. |
| | Seating arrangement, rules and techniques will remain the same as in the earlier session. |
| Distribute handout: | |
| VA-10 (remaining sheets) | |
| | Here are the remaining sheets on the case study on "Paper Reel Core". |
| | Hope they will provide you useful leads in your exercises. |
| Allow about one hour to complete the evaluation. | |

| Procedure | Statement |
|---|---|
| While the teams are in 'sessions', go round their tables to ensure that nobody is:<br><br>  – dominating or monopolising the proceedings;<br>         or<br>  – pushing his own ideas as 'the best';<br>         or<br>  – digressing from the subject or procedure.<br><br>When the groups reassemble, leaders present their 'cases' and spell out their choices from amongst the suggested ideas.<br><br>Thank them and return them to their seats.<br><br>Announce tea break. | |
| **CASE STUDIES IN FULL** ||
| Cover by Lecture Method:<br><br><br><br><br>Explain procedure in short.<br><br>(*depends on the strength of each team). | So far, we have practiced different Phases of the Job Plan, almost Independently.<br><br>Now, we will apply them collectively to the problems from your own departments or shops.<br><br>In each team there are (5) * members and each member must have identified at least one job. |

| Procedure | Statement |
|---|---|
| Selection of job by **Matrix Method:** | Thus, each team has at least (5)* jobs, One of them has been taken up earlier, leaving a balance of at least (4)* jobs. |
| | Each team will apply matrix method of evaluation to assess relative importance of the jobs, selecting appropriate criteria like potential savings, technology improvement serviceability, market demand, import substitution, etc. |
| | You can then apply the job-plan phases to your jobs, giving top priority to the 1st ranked job. |
| INFORMATION: | The 'problem owner' then discloses all the information to the other members of the team and clarifies the doubts. |
| | The team first defines the problem and then the functions(s). |
| SPECULATION: | The team conducts a brainstorming session, observing the procedure meticulously. |
| If possible, invite members from other teams to participate in the session. | Group leader lists all ideas. |
| EVALUATION: (The evaluation may preferably be done on the following day). | In evaluation, the team selects the criteria, gives weightages to them and then rates the effectiveness of various alternatives by allotting score points against the selected criteria. |

| Procedure | Statement |
|---|---|
| | The analysis finally leads to a decision suggestive of optimum solution.<br><br>That in brief is the procedure that we will follow in the remaining time/in the next session. |
| Announce "Home Work":<br>(Chart preparations) | Each team will prepare wall posters or charts, putting in nutshell, the information data, function-cost analysis, list of short-listed ideas, and matrices of evaluation up to decision level.<br><br>The charts will be exhibited on the wells, before the commencement of our next session. |
| – Clear doubts. | |
| – Ensure that the necessary stationery like drawing sheets, sketch pens, drawing instruments, etc. for making charts are available. | |
| – (Announce prizes for the best presentation.) | (The best presentation will be awarded a special prize.)<br><br>Good luck and thank you. |
| Close Session. | |

# 12

# VALUE ENGINEERING WORKSHOP

SESSION - XI

— PROGRAMME PLANNING &
EXECUTION PHASE.

— IMPLEMENTATION PHASE.

— FINAL REPORT PHASE.

# VALUE ENGINEERING WORKSHOP

## Session-XI: - PROGRAMME PLANNING & EXECUTION (i.e. DEVLOPMENT) PHASE.

### - IMPLEMENTATION PHASE.

### - FINAL REPORT PHASE.

| Procedure | Statement |
|---|---|
| **Make suitable opening remarks.** | |
| **PROGRAMME PLANNING & EXECUTION PHASE** | |
| Objectives:<br><br>– To point out that the evaluation phase helps in short-listing a very few but most promising ideas, but not in arriving at the final decision.<br><br>– To emphasise the importance of rigorous appraisal of the short-listed ideas, and final evaluation.<br><br>– To present a practical proposal which will stand up to the specialists' microscopic searches for defects. | |
| **BRIEF REVIEW OF EVALUATION** | |
| | We have seen that the evaluation phase consumes a lot of time and effort in selecting optimum solution or solutions.<br><br>We have also seen that the process of decision making involves choosing of suitable criteria, weightage scale and finally, application of an appropriate quantitative method to determine the relative effectiveness of the alternatives. |

| Procedure | Statement |
|---|---|
| | In simple case studies, like that of the pen holder, we have presumed prior knowledge of various attributes and factors as applied to various alternative ideas.

For example, we estimated the cost, durability and service life of plywood to be less than that of metal strip; the cost of chrome plating to be more than that of plastic coating; and so on.

The assumption based on repetitive experiences are alright in case of simple criteria.

But in Engineering and high technology fields, even the 'most promising' ideas will face the perils of presumption, if rigorous appraisal is not carried out before making final recommendation.

By rigorous appraisal, I mean developing the potential ideas into acceptable proposals. |
| **STEPS IN PLANNING & EXECUTION (DEVLOPMEMNT)** ||
| Summarised statement of steps in Planning & Execution phase: | This involves preparation of idea-investigation check list of each of the selected criteria, drafting a programme, discussion with specialists, suppliers, customers, and colleagues in other departments, etc.

Listing of an anticipated resistance to changes and charting of strategies to overcome the possible obstructions are also a part of this phase.

Prototypes, trial testing, collection of feed backs and finally, making recommendation of the most attractive alternative for implementation, are the other activities of this phase. |

| Procedure | Statement |
|---|---|
| | These activities clubbed together constitute out 5th phase: namely, Planning & Execution Phase. |
| <u>C.B.</u><br><br>   – Put a tick mark on "5: Programme Planning & Execution Phase," in 1st col.<br><br>   – Write the heading of this phase in 2nd column:<br><br>**5. <u>Planning & Execution Phase</u>**<br><br>\* Development of ideas and final recommendation.<br>   – Check lists of criteria Investigation.<br>   – Plan of action for:<br>     \* discussion<br>     \* overcoming 'road blocks'<br><br>   – Prototype making<br>   – Testing/proving<br>   – Feed back<br>   – Final recommendation | That in brief is the gist of our present session. |
| <u>Long pause to allow copying of points from C.B</u> | |
| **ELABORATION OF STEPS IN DEVELOPMENT OF IDEAS** | |
| The following statements (Up to page 182-Syndicate Work) are valid only if the booklet titled 'HAND OUTS ON VA/VE & CREATIVITY' is distributed.<br><br>Vide page 7<br><br>Check validity of the following:<br><br>(Exhaustive check-lists are already in hands of participants. cf. VE-15/VIII and VE-16/VIII. | |

| Procedure | Statement |
|---|---|
| Put a few questions on 'Check lists' of two or three criteria to maintain continuity on idea-development steps.) | The 1$^{st}$ step in programme planning is preparation of checklists.<br><br>What should include in the investigation check list for, say, product design? |
| Draw out 3 or 4 of the following:<br><br>– Performance function;<br><br>– Tolerances; | ( – what are the required tolerances?<br>– Can they be relaxed? |
| – Any special materials & process; | – Any special tooling or equipment required? |
| – Operating environment & life (fatigue, wear) | – Any corrosive or dusty atmosphere? |
| – Fail-safe features; | – Does the function call for duplication of safety provisions? |
| – Inter-changeability; | – What is the degree of inter-changeability of parts or sub-assemblies? |
| – Use of standard components; | – Can any standard parts be used?) |
| – Optimum mechanical simplicity;<br><br>– minimum weight | |

| Procedure | Statement |
|---|---|
| | What about quality checks? |
| Bring out few factors like:<br><br>  – Customer's requirement;<br><br>  – Measuring facilities;<br><br>  – Material testing;<br><br>  – Pressure testing;<br><br>  – Functional testing; etc | |
| | In addition to Design & Quality, and irrespective of the product, it is essential to prepare checklists of;<br><br>  – Performance;<br>  – Manufacturing;<br>  – Materials;<br>  – Marketing; etc. |
| \*   Make this statement only if the handouts VE-15/VIII and VE-16/VIII are distributed earlier. | |
| | \*  (You have been given two comprehensive checklists in 8th session.<br><br>I hope you could find time to go through those lists and select appropriate questions from them to compile the required checklists for your jobs.) |
| \*   (Pause to scan through the Checklists).<br><br>Step-2:<br><br>DRAFT A PLAN OF ACTION | |
| | Having prepared the checklists, we should ensure that all the questions are answered: |

| Procedure | Statement |
|---|---|
| <u>C.B.</u><br><br>(Copy from a pre-prepared card)<br><br>Who will help?<br>   – Designer?<br>   – Buyer?<br>   – Shop Manager?<br>   – Expert?<br><br>Any discussion?<br>   – With whom?<br>   – When?<br><br>Any additional data?<br>   – Who will collect?<br>   – When?<br><br>Any prototype/testing?<br>   – What cost?<br>   – Time?<br><br>_____<br><br>Read from C.B. | Who will answer?<br>   – Designer?<br>   – Buyer?<br>   – Supplier?<br>   – Any expert or specialist?<br><br>Any discussion called for?<br>   – with whom & when?<br><br>What additional information/data will be required?<br>   – Who will collect it?<br>   – When?<br>   – Any proving or prototype required?<br>   – If so, how long will it take to make it or prove it?<br>   – What will that cost?<br>   – Are the efforts, time & cost commensurable with the anticipated gains? |

| Procedure | Statement |
|---|---|
| | Practical experience of value engineers confirms that if the checklist is worked through question by question, no difficulty will be experienced in preparing a thoroughly sound and workable plan for further investigation in respect of each of the shortlisted alternatives ). |
| <u>Long pause</u> | |
| **SYNDICATE WORK** | |
| | You have your own 'jobs' fully analysed up to evaluation, with a few shortlisted alternatives to achieve each of the basic functions. |
| | I will now ask you to prepare <u>your plan of action for</u> developing the ideas. |
| | As before, we will work in syndicates. |
| | Before we start our work, let me repeat some of the useful points. |
| | When preparing the programme of action; |
| |    –  Take your own time; don't rush through the steps. |
| *  Applicable only if VE-15/VIII & VE-16/VIII were distributed. | (* –  Scan through the checklists distributed earlier) |
| |    –  Make a questionnaire on key points pertaining to each of the criteria selected for evaluation. |
| |    –  Cross-check to ensure that all essential elements are covered. |
| | Next, prepare a question-answer list: |
| Refer to C.B | |

| Procedure | Statement |
|---|---|
|  | Who will answer? Designer? Line Manager? Expert?<br><br>If he cannot be available, what then?<br><br>Any discussion required?<br><br>   – with whom? when?<br><br>Any addition data called for before discussion?<br><br>   – What is that?<br>   – Who will collect?<br>   – When?<br><br>Any proving required? What will that cost?) |
| Allow about 30 minutes for syndicate work:<br><br>Re-assemble: |  |
| Invite 1<sup>st</sup> leader to present his team's programme.<br><br>– Instruct him to be brief.<br>– Check understanding.<br>– Ask him to invite questions of clarification.<br>– Thank him & return to his seat in group.<br><br><br>One by one, call the other syndicate leaders to the head of the table and ask them to present their 'programmes'.<br><br>Follow the same procedure as in the 1<sup>st</sup> case. | Will the 1<sup>st</sup> team leader come over here and read out the programme of his group?<br><br><br><br><br><br>Mr. _____ and his team have made a very useful contribution to our practice of preparing a **plan of action**. |

| Procedure | Statement |
|---|---|
| At the end of presentation:<br><br>– Announce 'Tea' break. | |
| C.B.<br><br><u>Step-3:</u> **Execution of plan:**<br><br><u>Objective:</u> To secure co-operation in developing-ideas to a practical proposition.<br><br>(To overcome road blocks & win the first race). | Now, we come to execution part of this phase.<br><br>Arming oneself with a sound and workable programme of action will not lead to smooth sailing.<br><br>On way to investigation & development of selected ideas, you will face many difficulties.<br><br>Please remember, there are people waiting to shoot down your idea birds at every step.<br><br>Your skill in handling human factors will be on severe test, and success or failure of your attempts to bring about the desired change will largely depend on your tact and patient in overcoming the obstruction and securing help of the concerned people.<br><br>One of the steps in the action plan is "to meet and discuss with designers, buyers, vendors, etc."<br><br>Who will decide the time and venue for discussion? |
| Pause for the obvious answer "the person from who we seek help." | |

| Procedure | Statement |
|---|---|
| If the answer is "we," ask the opinion of the majority. | |
| | (Do you agree that we should decide the time and venue for the meeting? Will he cooperate?) |
| | Do you think that some support data should be given well before asking for an appointment to discuss your Proposals? |
| | or |
| | Do you fear that such data will only help him in digging deep into the weakness of your proposals? |
| If the answers are of conflicting types, allow short discussion till they reach an agreement in favour of giving Support data in advance. | |
| | What advantages will accrue by giving Support data in advance? |
| Pause for obvious response: | |
| "Allow time for consideration and preparing suggestions for improvement." | |
| **ROAD BLOCKS** | |
| | Having secured the appointment, what type of obstructions do you anticipate and why? |
| Bring out 5 or 6 of the common Mental blocks used as excuses to prevent change. | |
| * Statement is valid only if the handout VM-06/VII was distributed in Session-VII. | *(You may recall the 63 ways to prevent change.) |
| * (Allow them to go through the list for about 5 minutes.) | |

| Procedure | Statement |
|---|---|
| | Why do people resist changes? |
| Allow short discussion to draw out:<br><br>– attitude<br>– work habit<br>– fear<br>– laziness<br>– ignorance | |
| | People generally get used to doing things in certain way, in a certain place, at certain time, and with certain people. |
| | Now, every change has a smell of un-known danger or challenge to one's authority, expertise, position or status. Sometimes, it is shear laziness. |
| | How do you feel when somebody in your house changes something at home without telling you before hand? |
| "Usually resent it." | We are all inclined to question: Whether it is necessary to change the things to which we are accustomed. |
| C.B.: (3rd column)<br>Accept that resistance to change is normal. | |
| | One of the best way of overcoming resistance is to let the people know in advance changes that will affect them. If you could tell them WHY of it, it would be still easier to get them to accept the change. |
| Distribute VM-09/XI<br>(Little Red Hen)<br>Give 5-7 minutes to read.<br>No comments. | |

| Procedure | Statement |
|---|---|
| **ROLE PLAY** | |
| | Let us have a small session of Role Plays. I will play the role of the specialist, vendor or whoever you have planned to meet.<br><br>Will you please come to the top of the table and explain the setting? Please come and interview me. |
| – Invite one of the team leaders to the head of the table, with his 'job' (completed up to evaluation stage).<br><br>– Ask him to state the "setting" <u>briefly</u> i.e. the product or process, function, present cost, short-listed alternative ideas, and what he expects from you as a specialist, supplier or line manager.<br><br>– React fairly and naturally to his leads.<br><br>– Remember he has come to you for help and guidance.<br><br>Don't allow him to coach you. Stop him as soon as he attempts to do that.<br><br>– Vary your tone, pitch, gestures (twinkle in eye, e.g.), posture to reflect attitude of different specialists or line managers to resist change. | |
| | You know the reason of his meeting me, and have seen the approach he has adopted.<br><br>What do you say about it?<br><br>Has he achieved his objective?<br><br>Could the presentation be improved in any way? |

| Procedure | Statement |
|---|---|
| Discuss whether objective was achieved. If not, seek suggestions to improve the APPROACH.<br><br>At the end of discussion, thank and return the team leader to his seat.<br><br>   – Invite any other member from the same team to the head of the table and ask him to present the same "case" with suggestions incorporated.<br>   – Time permitting, conduct one or two more role-plays on members' jobs.<br>   – Summarise the discussions.<br><br><br><br><u>C.B.</u> (3rd column)<br>(copy from a pre-prepared card)<br><br>   – Ask for help;<br>   – Don't argue;<br>   – Listen;<br>   – Positive questioning;<br>   – Welcome suggestions;<br>   – Show respect to experience & professional standing;<br>   – Don't jump to conclusions;<br>   – Stand near the board and comment in 'matter-of-fact' tone. | Your intention of meeting specialists, line managers, vendors, etc. is "to ask for help in developing ideas" and not for imposing change.<br><br><br><br>The common sense approach is to present the case briefly and objectively and<br><br>Then to sit back and listen.<br><br>   – Don't argue. |

| Procedure | Statement |
|---|---|
| | Question in a positive form and respond positively to suggestions for improvement.<br><br>– Don't interrupt.<br><br>Don't do all the talking yourself.<br><br>Guarantee acknowledgement of contribution to improvement.<br><br>Furnish additional data asked for, as early as possible.<br><br>Ensure that person sitting opposite to you gets the feeling that he has contributed his efforts to the improvement.<br><br>Please, also remember he is a specialist and any attempt to teach him or coach him will only antagonise him. |
| Final choice of alternative: | After discussions with specialists and line manager comes the time for little introspection, re-investigation and final evaluation.<br><br>De-bug your proposals of lacunae, if any; collect data and compile comparative statement on the developed ideas.<br><br>The table should indicate as to how each developed idea compares with the existing design and with the other<br><br>– estimated costs;<br>– time and ease of implementation;<br>– effects on selected criteria, etc.<br><br>Based on our discussions so far, and taking some leads from the case studies on Paper Reel Core, etc. |

| Procedure | Statement |
|---|---|
| | you may now compile reports on your VE studies for presentation to the representatives of the management, in the concluding session of our deliberations. |
| – Invite questions of clarification.<br>– Clear doubts.<br>– Distribute copies of book-let "Guidelines for preparing V.E. reports"(by the same author), vide page No.7, Para.2.13<br>– Announce time & venue for presentation of reports by the team leaders.<br>– Clean C.B. (except Col.1) | |
| **IMPLEMENTATION PHASE** | |
| C.B.<br><br>Put a tick mark on:<br>"6. Implementation" in the 1st column, and write the heading in 2nd column. | Coming to the last two phases of the job plan, "implementation" and "final report," you are likely to face a tougher task in implementing your proposals.<br><br>As we have discussed earlier, any change will affect a large number of people or departments to varying degrees.<br><br>For example, if a change in design is called for by your proposals, many things will have to be done:-<br><br>– preparation of fresh drawings;<br><br>– Replacement of old drawings with new drawings; |

| Procedure | Statement |
|---|---|
| | – Ordering of new materials, tools, jigs & fixtures, gauges, etc. |
| | – Arranging of test facilities; |
| | – Training of operators, etc. |
| | Concept of packing or packaging, shipping and marketing may also undergo complete change. |
| Invite comments from the group on likely problems of implementation | |
| | What problems do you foresee in Implementing your proposals? |
| Draw out factors relating to: | |
| – Design, material specifications & procurement; | |
| – Manufacturing operations (tooling, J&F, Training); | |
| – Inspection/Q.C.; | |
| – Testing (proving); | |
| – Marketing (packaging, dispatch); | |
| – Field service; | |
| – Maintenance; etc. | |
| | **Important Notes:** |
| | To conclude our discussion, let me add that people have tendency to fall back in the old track. |
| | Implementation needs close follow-up till the change becomes a routine practice. Then only can one be sure that the new ideas have been fully implemented. |

| Procedure | Statement |
|---|---|
| **FINAL REPORT PHASE** ||
| <u>C.B.</u><br><br>Put a tick mark on "7. Final Report" in the first column.<br><br>*Applicable only if the following booklet is distributed.<br><br>(Refer to "Guidelines for Writing VE/VM report".)<br>-Page 5, Para. 2.13 | * (Coming to the last phase i.e. "preparation of final report" on your VA/VE project, enough is said in the booklet just distributed.)<br><br>I would like to reiterate that the final report should reflect the value of your efforts, and contribute to the common fund of knowledge for the benefit of others in the line, now and in future. |
| Invite questions and comments on topics covered in this workshop | Before concluding my part in this workshop, will you please help me in improving this V.E. programme, by offering your suggestions?<br><br>You are also welcome to ask questions. |
| Answer the questions enthusiastically and make note of the suggestions, conspicuously.<br><br>Ensure that the workshop ends with a pleasant feelings.<br><br>Acknowledge that you have learnt a lot from them and has a pleasant time in their company.<br><br>  – Announce your intentions to keep contacts with them till the projects are completed. | |

| Procedure | Statement |
|---|---|
| **DETAILS OF SEMINAR** ||
| Announce details of Seminar: | |
|     – The name and designation of the Chief Guest and also of other dignitaries. | |
|     – Time. | |
|     – Venue. | |
|     – Sequence of presenting Papers in the seminar. | |
|     – Prize distribution (if any). | |
| Request organiser of this workshop to distribute invitation cards amongst the participants. | |
| After this Session: | |
| Ensure that the two eminent persons (judges) who are invited by the Organiser of workshop, are available for evaluation of the "Presentation of Reports". | |
| Also, meet the 'Chief Guest' and apprise him of the workshop deliberations. | |
| Supply key points to him. | |
| (If possible, provide a draft for his address.) | |
| CLOSE SESSION. | |
| CLEAN C.B. (completely) | |

# 13

# VALUE ENGINEERING WORKSHOP

## Session - XII

## — SEMINAR ON VE

# VALUE ENGINEERING WORKSHOP

## Session-XII:- SEMINAR ON VE

| Procedure | Statement |
|---|---|
| – Invitation to Heads of department, top executives (3 of them as judges).<br><br>– Spend 10 minutes with each group to understand what & how they would like to present? What aids? Who is going to present?<br><br>– Explain the sequence of presentation.<br><br>– Announce procedure:<br><br>  Who is presenting?<br>  What he should do?<br>  Time limit.<br><br>                   (Tea/Coffee) | |
| – welcome address by the Trainer.<br><br>– Address by Chief Guest.<br><br>(Distribute copies of address immediately after presentation is over.)<br><br>– Announce prizes.<br><br>– Prize distribution.<br><br>– Vote of Thanks<br>(*Script to be prepared earlier). | |
| Evening:    High Tea or Dinner<br>                With Guests & Spouses. | |

# 14

## APPENDICES

# APPENDIX - I

## LIST OF FLIP CHARTS OR TRANSPARACIES FOR OVERHEAD
## PROJECTION OR SLIDES

| S.NO. | Chart No./Session No. | Description |
|-------|------------------------|-------------|
| 1. | FC-01/I | Objective of VE/VM work shop |
| 2. | FC-02/I | Economic Value |
| 3. | FC-03/I | Refills of Jotter Pen |
| 4. | FC-04/I | Graph: Quality Vs Cost |
| 5. | FC-05/I | Classes of Value |
| 6. | FC-06/I | Definition of VA/VE |
| 7. | FC-07/I | VA/VE : A multi purpose tool |
| 8. | FC-08/I | 33 KV MOCB/3D Sketch |
| 9. | FC-09/II | Schematic of Circuit Breaker |
| 10. | FC-10/II | Flange |
| 11. | FC-11/II | Insulators |
| 12. | FC-12/IV | Total business curve |
| 13. | FC-13/IV | Product Life Cycle |
| 14. | FC-14/IV | Cost Structure |
| 15 | FC-15/V | VA/VE Saving potentials |

| APPENDIX-I (1) |
| --- |
| FC-01/I |

<u>VALUE ENGINEERING</u>

<u>PROGRAMME OBJECTIVE:</u>

TO DEVLOP SKILL IN ACHIEVING

NECESSARY FUNCTION OF A

PRODUCT OR SERVICE FOR

A MINIMUM COST WITHOUT

DETRIMENT TO QUALITY

RELIABILITY AND PERFORMANCE.

| | APPENDIX-I (2) |
|---|---|
| | FC-02/I |

## VALUE ENGINEERING

VALUE:     ECONOMIC VALUE

- DESIRABILITY
- SCARCITY

IT VARIES WITH

- PERFORMANCE

- PLACE

- TIME

- QUANTITY

※     VALUE IS THE LEAST COST THAT
CAN ACCOMPLISH RELIABLY A
FUNCTION OR SERVICE

$$VALUE = \frac{WORTH\ (TO\ YOU)}{COST\ (YOU\ PAY)}$$

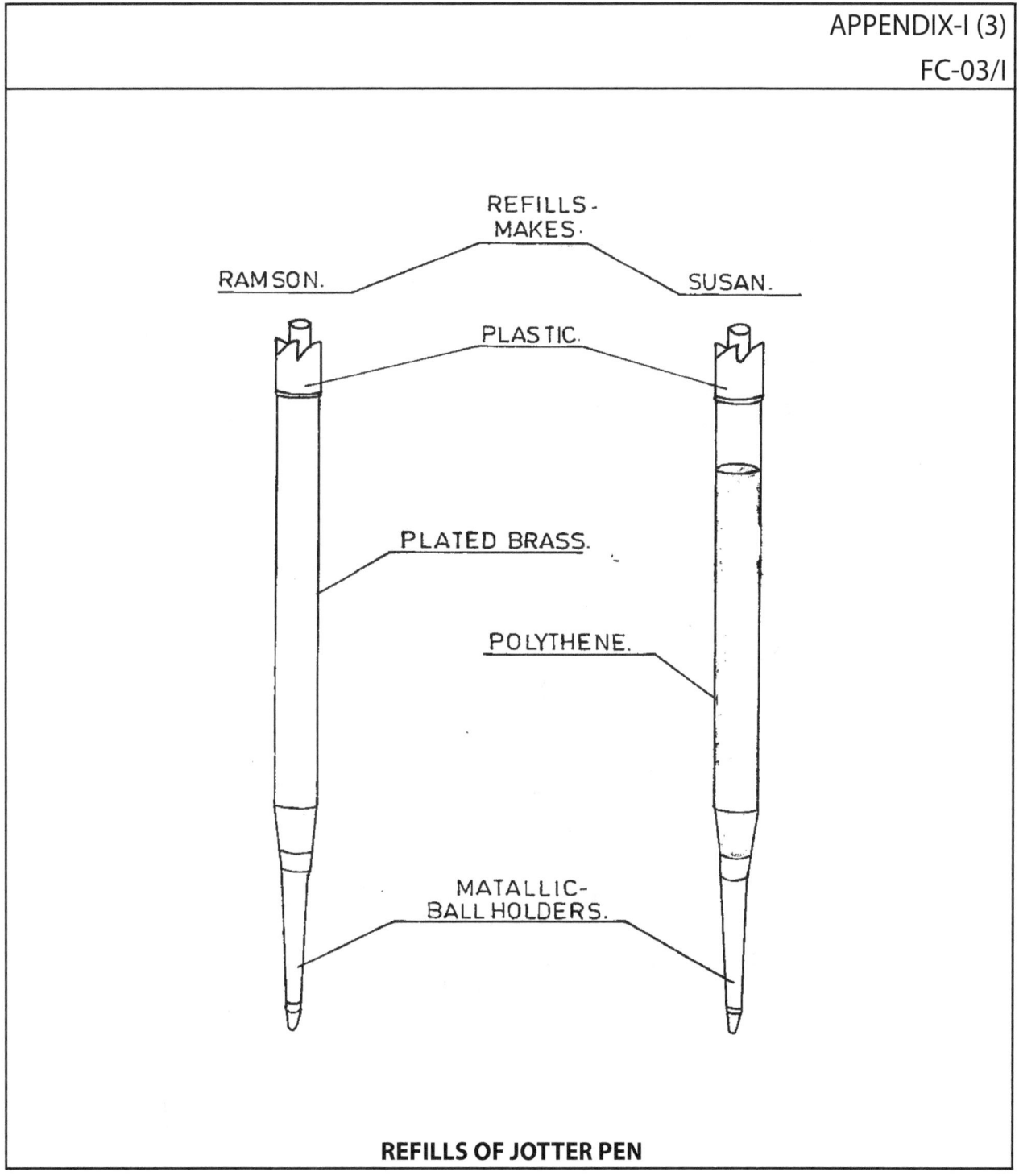

**REFILLS OF JOTTER PEN**

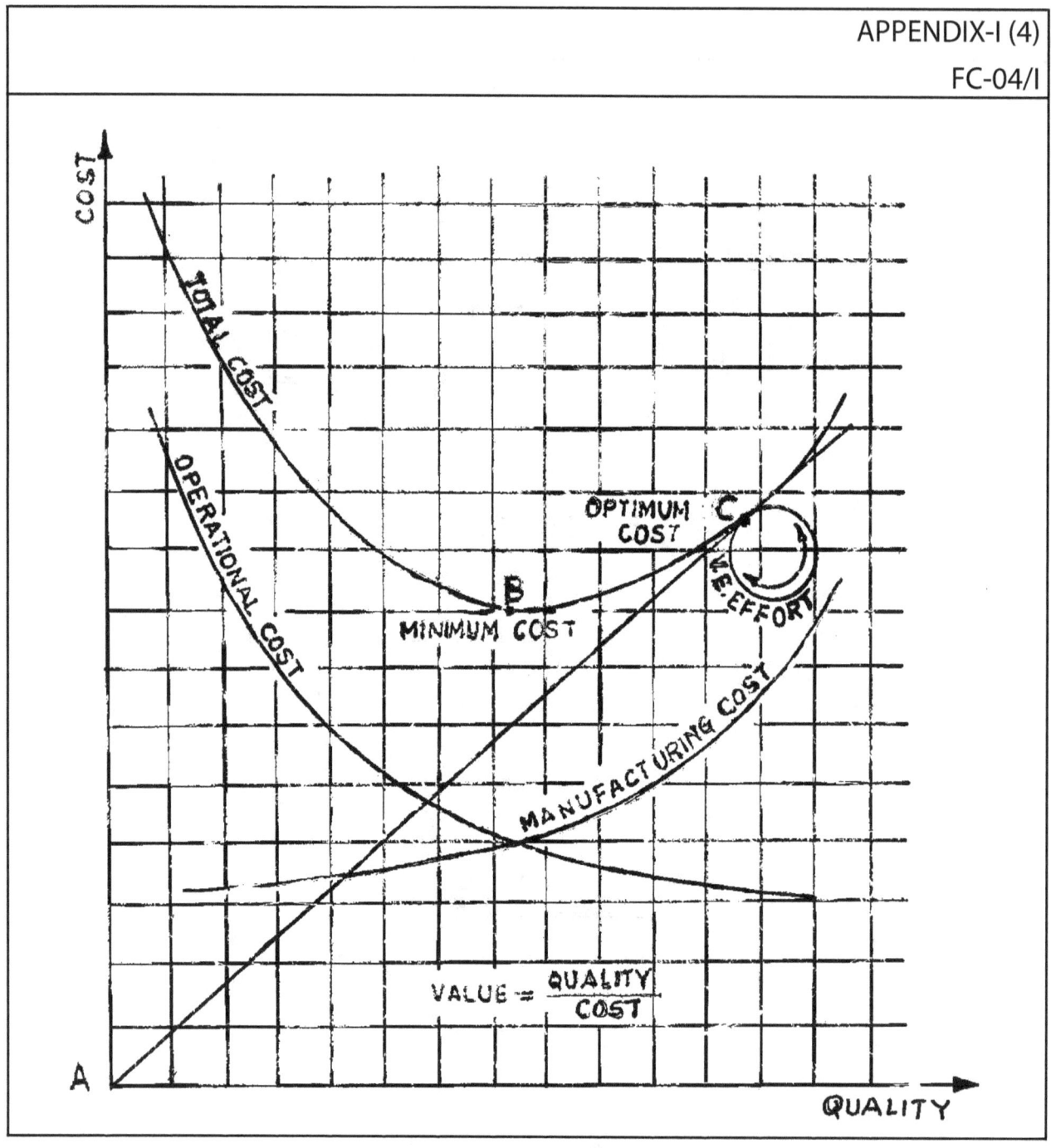

# VALUE ENGINEERING

## CLASSES OF VALUE

1. USE VALUE }           PHYSICAL
2. ESTEEM VALUE }        ATTRIBUTES

3. EXCHANGE VALUE }      ECONOMIC
4. COST VALUE }          CONSIDERATIONS

## TOTAL VALUE

**ENGG PRODUCT**                    **PAINTING/IDOL**

NB: You may add colours to the sectors

APPENDIX-I (6)

FC-06/I

## VALUE ANALYSIS/VALUE ENGINEERING

DEFINATION

VA/VE IS AN ORGANISED
METHOD OF IDENTIFYING &
ELIMINATING UNNECESSARY
COSTS-

OR

A DISCIPLINED PROCEDURE
DIRECTED TOWARDS THE
ACHIEVEMENT OF NECESSARY
FUNCTIONS FOR MINIMUM COST

WITHOUT DETRIMENT TO QUALITY
RELIABILITY & PERFORMANCE

R.G. CHAUDHARI

# VA.VE  A  MULTI PURPOSE TOOL

COST REDUCTION

INVENTORY
CONTROL

TECHNOLOGY
UP GRADATION

STANDARDIZATION

QUALITY
IMPROVEMENT

IMPORT
SUBSTITUTION

HIGHER PRODUCTIVITY

EXPORT PROMOTION

TEAM SPIRIT

205

APPENDIX-I (8)

FC-08/II

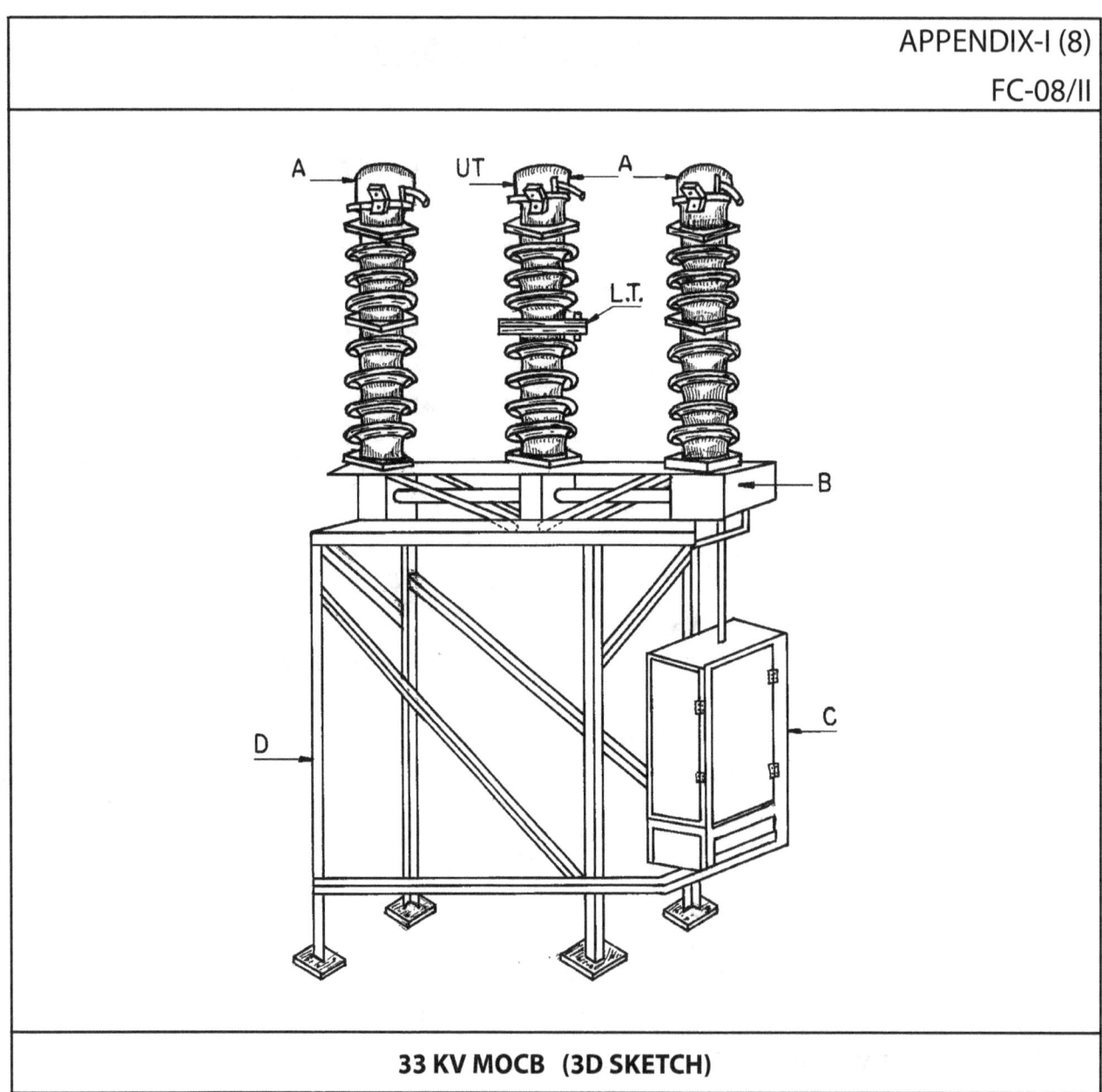

**33 KV MOCB  (3D SKETCH)**

| | APPENDIX-I (9) |
|---|---|
| | FC-09/II |

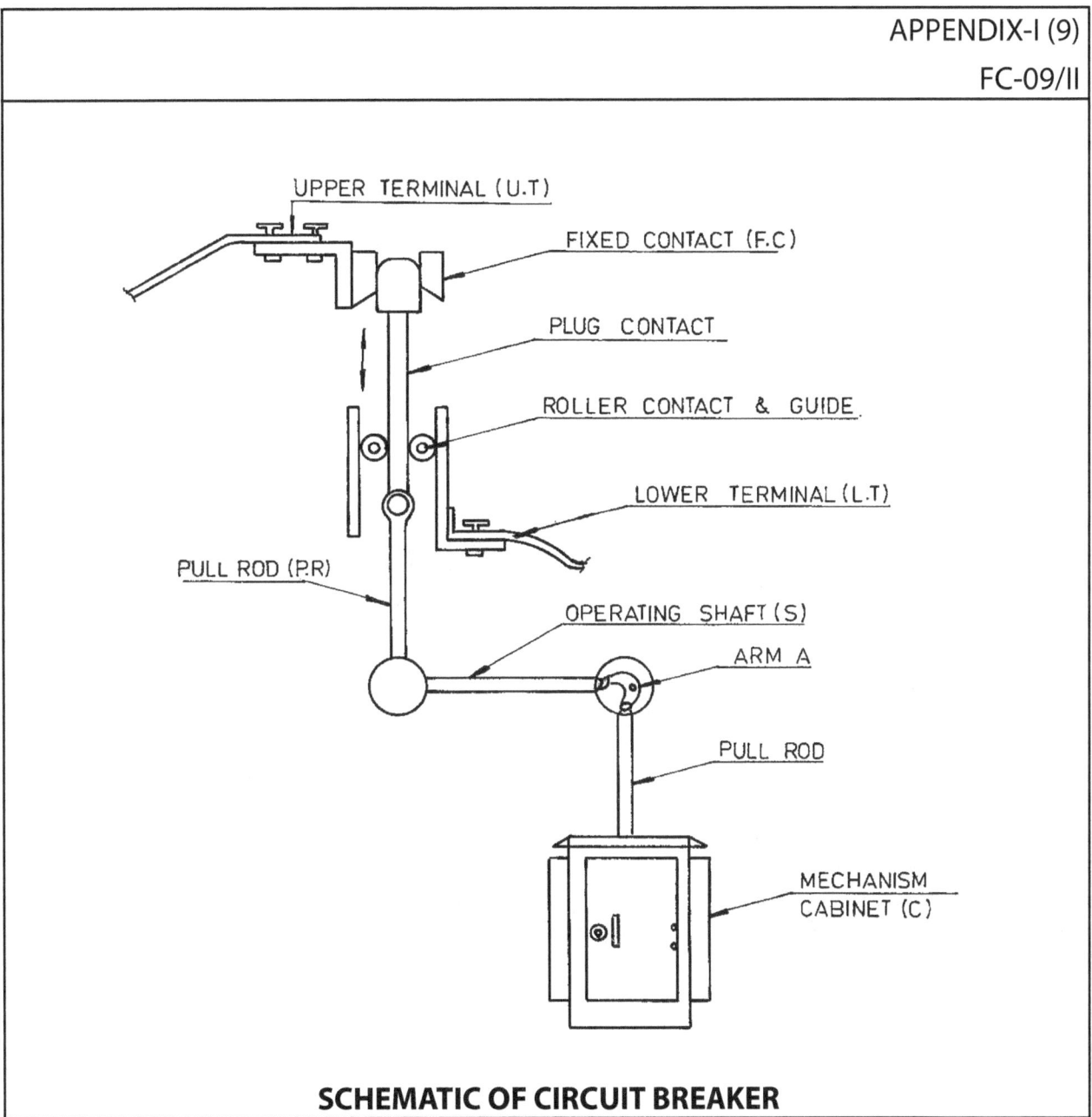

**SCHEMATIC OF CIRCUIT BREAKER**

**This is also <u>VA-02/II</u>**

**Page 5 of 10**

APPENDIX-I (10)

FC-10/II

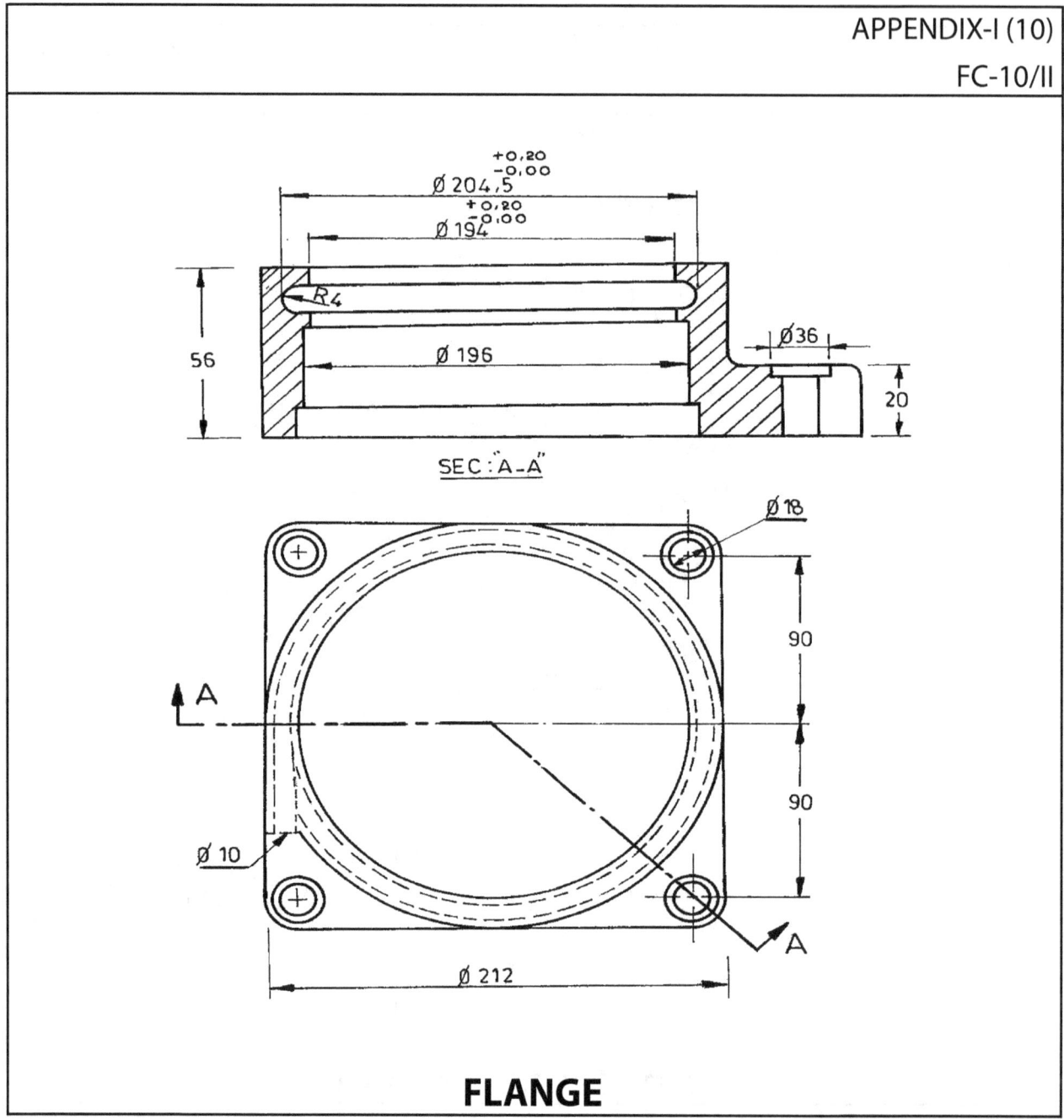

**FLANGE**

APPENDIX-I (11)

FC-11/II

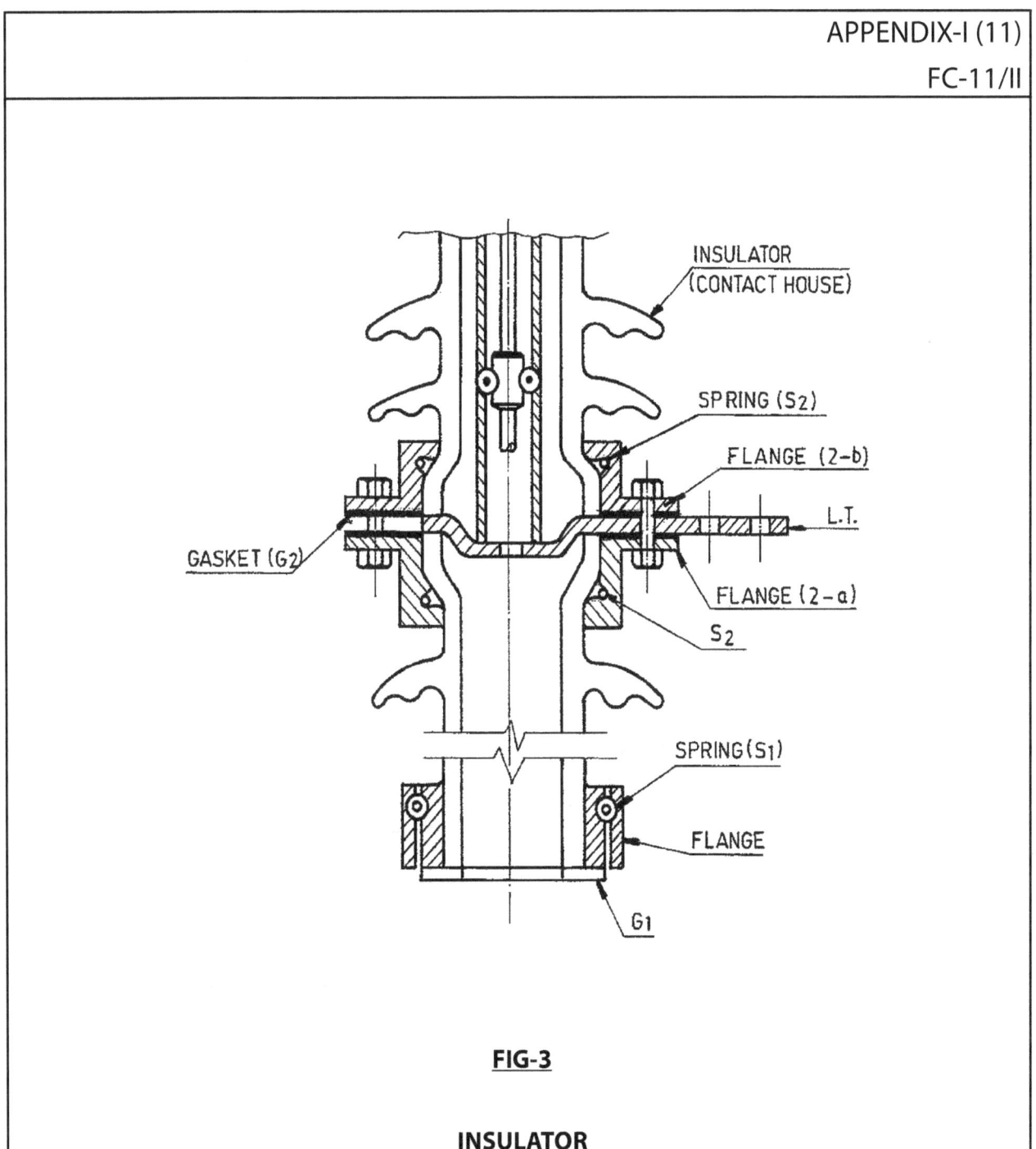

**FIG-3**

**INSULATOR**

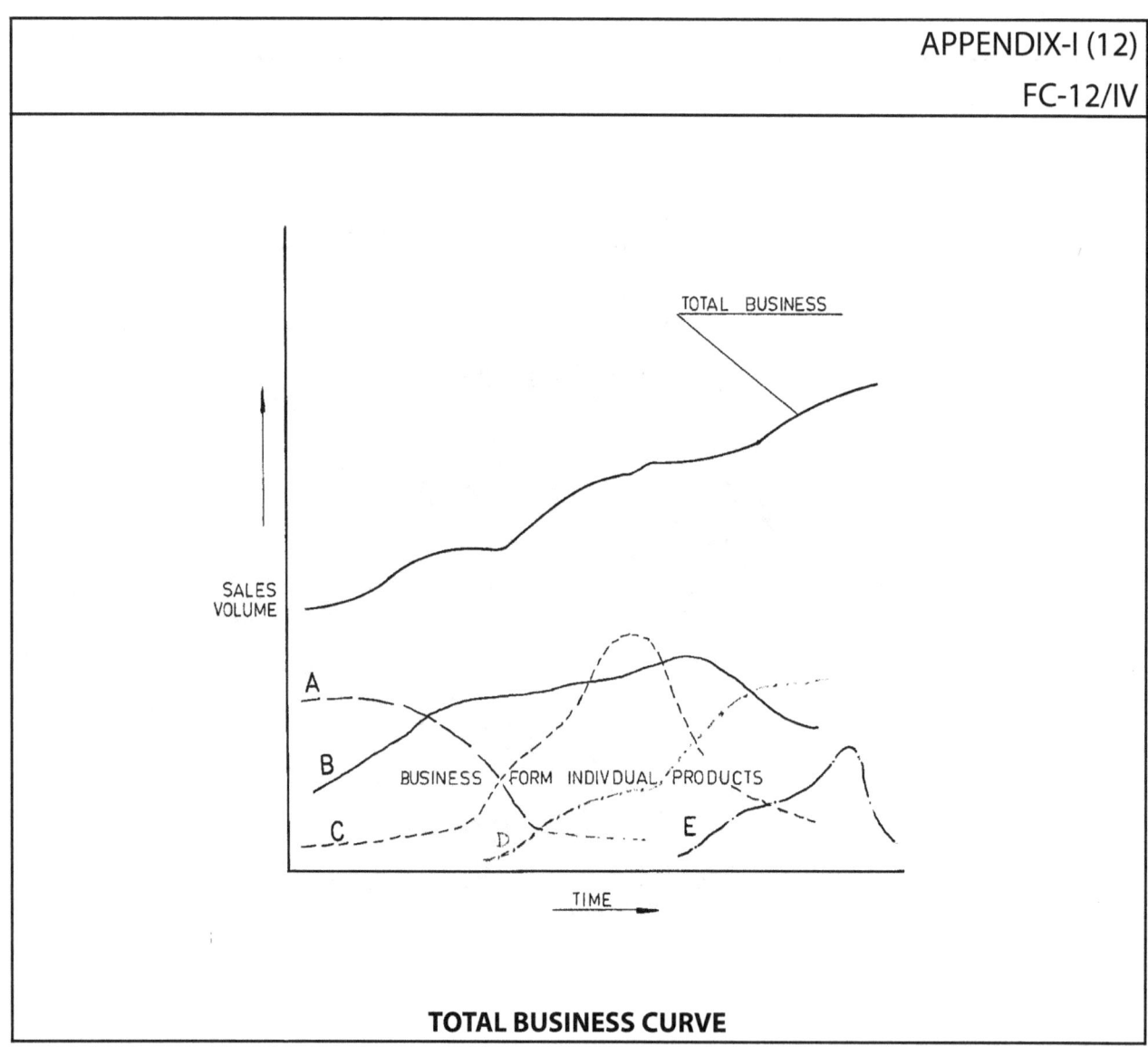

**TOTAL BUSINESS CURVE**

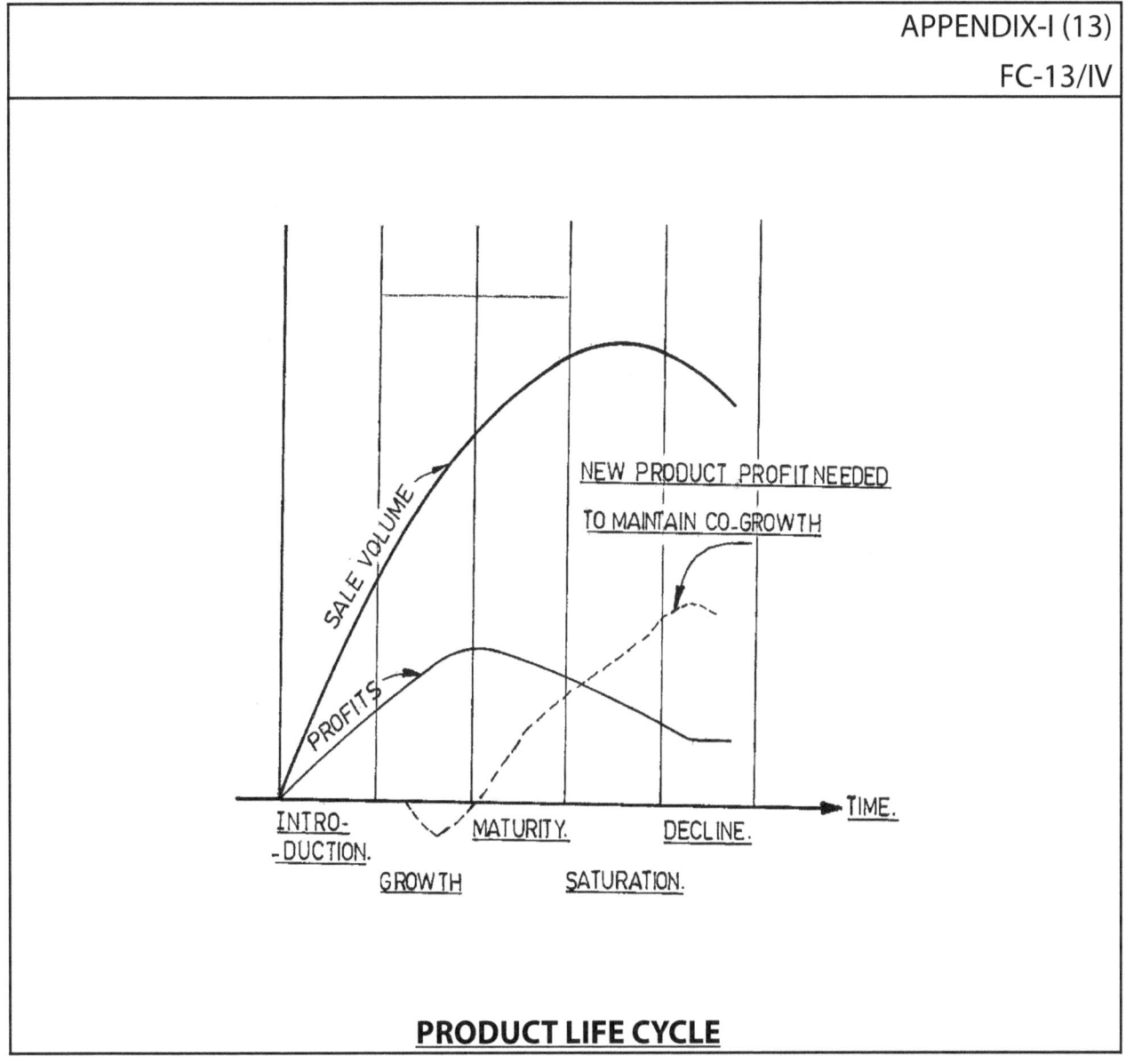

**PRODUCT LIFE CYCLE**

## THE CLASSICAL GENERALISED COST STRUCTURE

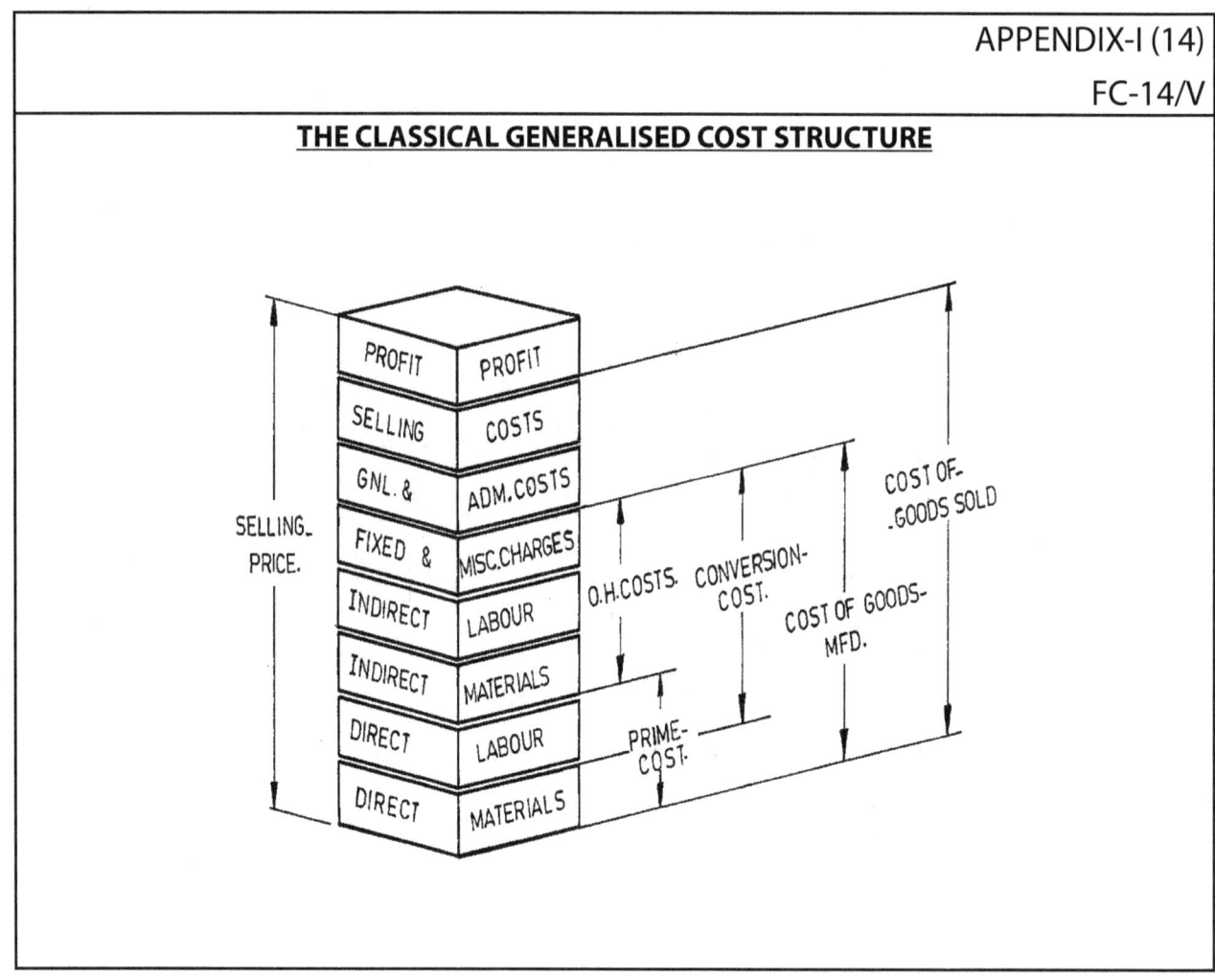

R.G. CHAUDHARI

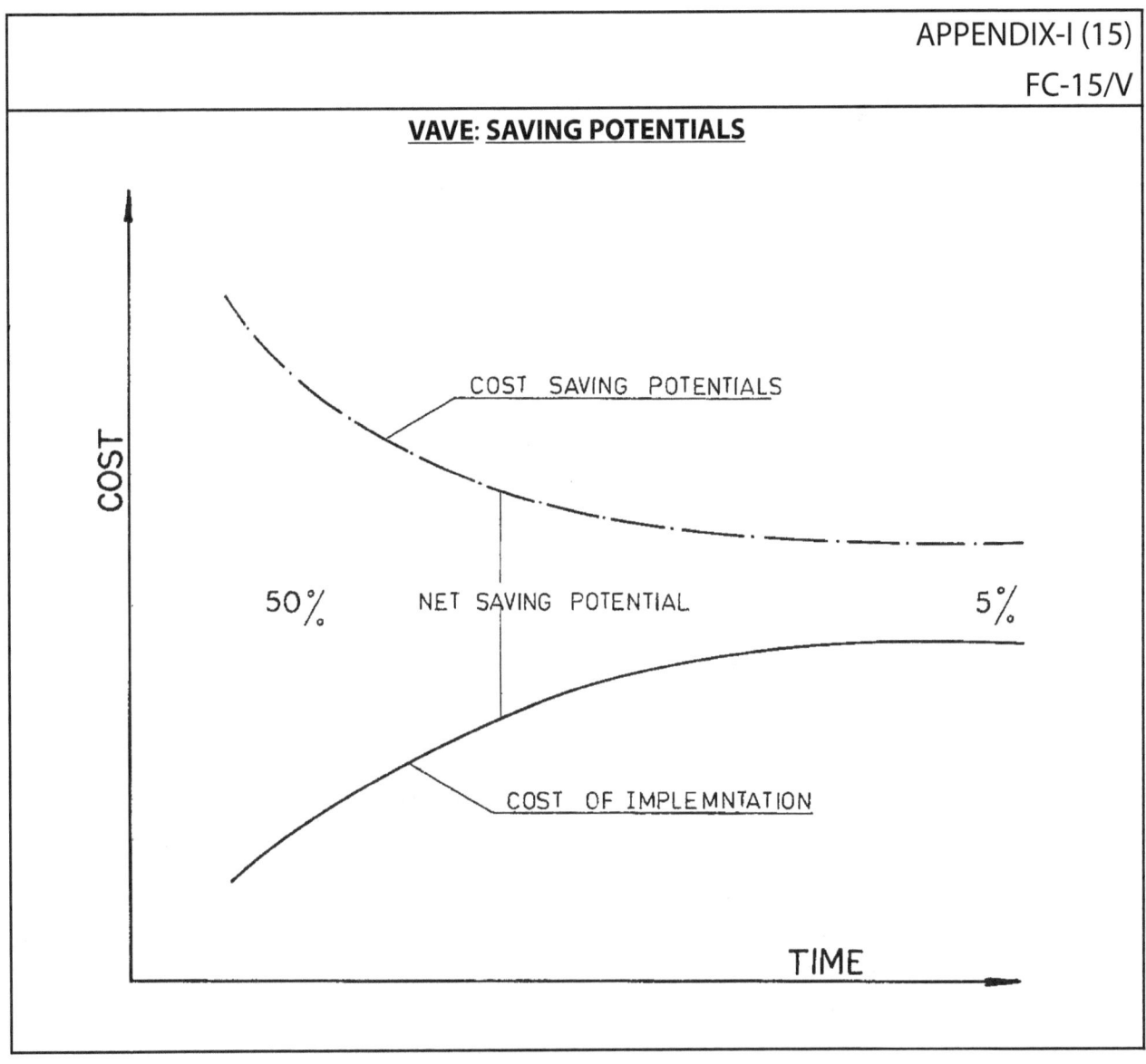

**VAVE: SAVING POTENTIALS**

COST SAVING POTENTIALS

50%    NET SAVING POTENTIAL    5%

COST OF IMPLEMNTATION

COST

TIME

# LIST OF VE/VM HANDOUTS FORMING PART OF COURSE MATERIAL

| SI No | Handout No./ Session No. | Description | No. of pages |
|-------|--------------------------|-------------|--------------|
| 1. | VA-01/I | Causes of Unnecessary Costs | 1 |
| 2. | VA-02/II | Job Plan Phases : Thru' A Case Study | 10 |
| 3. | VA-03/II | Job Plan Phases | 2 |
| 4. | VA-o4/IV | Golden Case Limited – A Case Study | 4 |
| 5. | VA-05/V | Information/Data Collection | 3 |
| 6. | VA-06/VI | Definition of Function : Verb + Noun | 2 |
| 7. | VA-07/VI | Functional Analysis – Ink Pot | 1 |
| 8. | VA-08/VI | Functional Cost Analysis : 2– pin plug | 1 |
| 9. | VA-08/VI | Functional Cost Analysis : 2 – Pin plug (Trainer's Copy) | 1 |
| 10. | VA-09/VI | Function Levels (General) | 1 |
| 11. | VA-10/VIII | Core of paper Reel -A case Study | 9 |
| 12. | VA-12/IX | Design of a Pen Holder – A Case Study | 2 |

# CAUSES OF UNNECESSARY COSTS

1.  MANAGEMENT  INEFFICIENCY:
    *   Complacency:(Routine cost reduction schemes will take care of unnecessary cost.
        Then why VE?)
    *   Lack of Value Objectives.
    *   Lack of Planning.
    *   Lack of Pressure.
    *   Lack of knowledge.

2.  INABILITY  TO  APPLY  VAVE:
    *   Lack of (Technical and cost) information.
        -   Full and accurate information helps in taking meaningful decisions.
        -   Information has to be continuously up-dated.
        -   Lack of information is the source of a false sense of security.
    *   Lack of communication.
    *   Lack of good ideas.
        -   Available talent is seldom fully utilized.
        -   (Many good ideas are still-born because they are never recorded or developed).
    *   Lack of Training.

3.  HUMAN WEAKNESSES:
    *   Honest wrong beliefs.
        -   Harmless ones like the 'word is flat'.
        -   Harmful and costly: like 'titanium can't be cast'.
    *   Habits and Attitudes.
        -   Relectance to accept new ideas.
        -   Obstructive and negative attitudes.
        -   Duplicating safety features.
        -   Finer tolerances than required.
        -   Excessive packaging.
        -   Over design, etc.
    *   Pride in own ideas.

4.  COMPETITIVE PRESSURES:
    -   Anxiety to beat competition
    -   Lack of time
    -   Over confidence

Result : Complicate/
Over design and
high costs.

# JOB PLAN PHASES : THRU' A CASE STUDY-1

## INFORMATION PHASE

### TABLE-I : BREAK-UP OF MATERIAL COST PER PRODUCT

(Rounded off to the nearest Rupee)

| Sl. No | Material | Cost in Rs. | | | % to Total Cost |
|---|---|---|---|---|---|
| | | Assy.1 | Assy.2 | Total | |
| 1. | Raw-Materials: | | | | |
| | (Sheets, Flats, Rounds, Tubes, etc.) | | | | |
| | (a) Steel | 1150 | 605 | 1755 | 9.5 |
| | (b) Non-ferrous metal | 1437 | 58 | 1495 | 8.0 |
| 2. | Insulation | 319 | 1 | 320 | 2.0 |
| 3. | Castings:* | | | | |
| | (Ferrous & Non-ferrous) | 4722 | 1399 | 6061 | 33.0 |
| 4. | Bought out items: | | | | |
| | (a) Imported | 2000 | 80 | 2080 | 11.0 |
| | (b) Indigenous | 2600 | 1200 | 3800 | 20.5 |
| 5. | Hardware | 801 | 512 | 1313 | 7.0 |
| 6. | Contingency (about 10%) | 1301 | 375 | 1676 | 9.0 |
| | Total: | 14,330 | 4,170 | 18,500 | 100% |
| *No: | (a) Bought out Castings | 480 | 829 | 1,309 | 7.0 |
| | (b) Internal Castings | 4,242 | 510 | 4,752 | 26.0 |
| | Total: | 4,722 | 1,399 | 6,061 | 33.0 |

# CASE STUDY-1 : INFORMATION PHASE

TABLE -2: COST DETAILS OF CASTINGS

(Rounded off to the nearest Rupee)

| Sl. No | Description | Material | Pieces/ Product | Cost Per Piece | Cost Per Product |
|---|---|---|---|---|---|
| 1. | Cylinder | Si. brass | 3 | 425 | 1,275 |
| 2. | Fixing flanges | Aluminium | 12 | 57 | 684 |
| 3. | Connection flanges | Aluminium | 3 | 213 | 639 |
| 4. | Connection flanges | Aluminium | 3 | 191 | 573 |
| 5. | Contact holder | Bronze | 3 | 110 | 330 |
| 6. | Worm rim | Brass | 1 | 322 | 322 |
| 7 | Top cover | Aluminium | 3 | 93 | 279 |
| 8. | Connection flanges | Aluminium | 3 | 71 | 213 |
| 9. | Frame | Steel | 1 | 141 | 141 |
| 10 | base | Aluminium | 3 | 34 | 102 |
| 11 to 17 | 7 items of which cost ranges from Rs.61/- to Rs. 100/- Per Product | | | | |
| 18 to 36 | 18 items of which cost ranges from Rs.41 to Rs.60/- Per Product | 1,503 | | | 1,503 |
| 37 to 60 | 25 items of which cost is less than Rs.20/-Per Product. | | | | |
| | | | | Total: | Rs.6,061 |

NOTES: VA-02/II: Page 3 of 10  is the same as Page 205

VA-02/II: Page 4 of 10  is the same as Page 206

VA-02/II: Page 5 of 10  is the same as Page 207

VA-02/II: Page 6 of 10  is the same as Page 208

# CASE STUDY-1: INFORMATION PHASE

## TABLE-3 : COST-FUNCTION MATRIX (BEFORE VA)

### BASIC FUNCTION TO FIX AN INSULATOR TO ANOTHER INSULATOR OR TO BASE FRAME.

| Sl. No | Elements of cost | Support weight (of insulators) | Hold Flange (to insulator) | Facilitate fixing (to base frame or another insulator). | Facilitate alignment | Resist corrosion | Shape parts | Total |
|---|---|---|---|---|---|---|---|---|
| | | F1 | F2 | F3 | F4 | F5 | F6 | |
| 1. | Casting (LM-6) | 28.00 | ... | 22.00 | 3.00 | 4.00 | | 57.00 |
| 2. | Phosphor Bronze Spiral Spring (imported) | .... | 16.00 | ... | 2.00 | 1.00 | .. | 19.00 |
| 3. | Chamfered gasket | .... | .... | 1.50 | .... | 1.50 | ... | 3.00 |
| 4. | Glue for item (3) | .... | .... | 0.15 | ..... | 0.15 | .... | 0.30 |
| 5. | Nuts & bottle (4 pairs) | .... | .... | 6.00 | .... | ... | ... | 6.00 |
| 6. | Machining of casting | .... | 3.00 | 3.00 | ... | 1.20 | ... | 7.20 |
| 7. | Grinding of insulator ends | ... | 3.50 | ... | ... | .... | ... | 3.50 |
| 8. | Assembly | 0.60 | 1.00 | 1.20 | 0.80 | .... | ... | 3.60 |
| 9 | Tooling | .... | ... | ... | ... | .... | 0.40 | 0.40 |
| | Total | 28.60 | 23.50 | 33.85 | 5.80 | 7.85 | 0.40 | 100.00 |

# SPECULATION PHASE

| S.No | Tentative ideas: |
|------|------------------|
| 1 | Use fabricated flanges |
| 2 | Form flanges on porcelain insulators and paste them together with epoxy. |
| 3 | Use simple pipe coupling and epoxy (adhesive). |
| 4 | Use G.I. casting and cement it to the porcelain end. |
| 5 | Use single piece insulator of longer length. |
| 6 | Change design of internal parts so as to accommodate them in a single insulator of existing size |

# CASE STUDY-1 : EVALUATION PHASE

### TABLE-4 –SIMPLE DECISION TABLE

| Sl No | Idea | Good points | Bad points | Decision |
|---|---|---|---|---|
| 1 | Use fabricated flanges | -less cost | - more machining<br>- more anticorrosion cost | Hold |
| 2. | Form flanges on insulators and join with epoxy based adhesive. | - less assy. time.<br>- less assy. matetials.<br>- imported spiral spring eliminated.<br>- grinding for insulator end faces eliminated. | - very costly.<br>- process not developed.<br>- outside business line.<br>- less strength.<br>- more damages.<br>- great difficulty in assy. of internal parts. | Reject |
| 3. | Use simple pipe coupling and adhesive. | - less cost (of pipe).<br>- imported spiral spring eliminated.<br>- grinding of insulator end faces eliminated. | - more cost of adhesive.<br>- change in orientation of lower terminal not possible.<br>- difficult to assemble internal parts. | Reject |
| 4. | Use G.I. casting and cement to insulator. | - less cost.<br>- better creep resistance.<br>- less machining & less parts.<br>- imported spring eliminated.<br>- grinding of insulator faces eliminated.<br>- less assembly time. | - more weight. | Accept |

# CASE STUDY-: PROGRESS PLANNING & EXECUTION PHASE

### TABLE-5: COST –FUNCTION MATRIX (AFTER VA)

Basic functions : to fix an insulator to another insulator or to the base frame

| SI No | Elements of cost | F1 | F2 | F3 | F4 | F5 | F6 | Total cost | | Saving |
|---|---|---|---|---|---|---|---|---|---|---|
| | | | | | | | | After VA | Before VA | |
| 1. | Casting | 11.00 | -- | 9.00 | 1.20 | 1.50 | -- | 22.70 | 57.00 | 34.30 |
| 2. | Cementing | -- | 9.00 | -- | -- | 1.00 | -- | 10.00 | 19.00 | 9.00 |
| 3. | Chamfered Gasket | -- | -- | 1.50 | -- | 1.50 | -- | 3.00 | 3.00 | -- |
| 4. | Glue for item(3) | -- | -- | 0.15 | -- | 0.15 | -- | 0.30 | 0.30 | -- |
| 5. | Nuts & bolts (4 pairs) | -- | -- | 6.00 | -- | -- | -- | 6.00 | 6.00 | -- |
| 6. | Machining of casting | -- | 2.00 | 2.00 | -- | -- | -- | 4.00 | 7.20 | 3.20 |
| 7 | Fixture of cementing | -- | 1.40 | -- | -- | -- | -- | 1.40 | 3.50 | 2.10 |
| 8. | Assembly | 0.60 | 1.00 | 0.20 | -- | -- | -- | 1.80 | 3.60 | 1.80 |
| 9. | Tooling | -- | -- | -- | -- | -- | -- | -- | 0.40 | 0.40 |
| 10. | Additional transportation cost (heavy cost) | -- | -- | -- | -- | -- | -- | -- | -- | -- |
| | Total: | | | | | | | 49.20 | 100.00 | 50.80 |

Annual Saving = Saving/Flange x No. of flanges per Breaker x No. of Breakers per annum.

= 50.80  x  12 x  300 = Rs. 1.82 lakhs.

Say: Rs. 1.80 lakhs.

# JOB PLAN PHASES

## A PRACTICAL PLAN FOR EFFICIENT IDENITIFICATION AND ELIMINATION OF UNNECESSARY COSTS WITHOUT DETERIMENT TO QUALITY AND RELIABILITY

| Phases | Objectives |
|---|---|
| I. ORIENTATION PHASE. | To identify and select right type of projects |
| | - sales analysis |
| | - Product analysis |
| | - cost analysis |
| | - new products |
| | - Field services/In-service reports. |
| | To define general scope, constraints and approach of study. |
| II. INFORMATION PHASE: | To obtain facts: |
| | - Marketing/Commercial. |
| | - Design Specifications and dimensioned Drawings |
| | - Operating conditions and environment. |
| | - Process and technological data. |
| | - Procurement, storage and handling |
| | - Product cost (material, labour, overhead, etc.) |
| | To define part & its Functions: |
| | - What is it? (Name Should indicate correct function performed) |
| | - What does it do? (Answer reveals function(s) performed) |
| | - Categorise function as primary, secondary, and unnecessary. |
| | - Specify Function by using a verb and a noun. |
| | To assess the degree of value opportunity: |
| | - Determine cost of function. |
| | - Allot a Cost/worth to each function. |
| | - Difference of 'cost' and 'cost-worth' is value opportunity. |
| III. SPECULATIVE OR CREATIVE PHASE: | To produce a no. of ideas for achieving the desired function. |
| | - Removal of barriers/road blocks. |
| | - Differed judgement/no criticism. |
| | - The more the better/quantity breads quality. |
| | - Combination and improvement of ideas. |

IV. EVALUATION PHASE:

To select the best ideas:

- Simple weighing
- Criteria T-charts.
- Criteria value chart.
- Forced decision

V. PROGRAMME PLANNING & EXECUTION PHASE:

To work out strategy for converting tentative ideas into tangible proposals and to executive the work plan.

– Ideas investigation check list and plan of action.
– Worth and cost of developing, proving.
– Discipline involved – mechanical, electrical, chemical, etc.
– Main functions (of each area).
– Workable plan.
– Discussion with specialists.
– Identifying and overcoming road blocks.
– Final selection of developed ideas.
– Gaining acceptance.
– Submission of report to Line Management.

VI. IMPLEMENTATION PHASE:

To actively follow up implementation of the proposals.

– Check lists for design, materials, manufacturing, in-service and quality.

VII. FINAL REPORT PHASE:

To record:

– Advantages gained.
– Ideas held back due to temporary constraints.
– Review dates for the ideas.
– Acknowledgement of help received.

# GOLDEN CASE LIMITED

– By R.G. Chaudhari

## A word about the Company:

The Company manufactures a wide range of products from a toaster to turbine and uses latest technology including CAM/CNC machines. Its manufacturing divisions include a captive Grey Cast Iron Foundry. It has recently diversified its activities in the field of solar energy. The real strength of the Company lies in its human resources. It employees about 12,000 people, of whom about 4,500 are highly skilled artisans and 1,200 Engineers & Technicians.

It is managed by competent professionals with progressive outlook.

During the proceeding five years, the output of the Company has increased from Rs. 6,504 lakhs to 11,425 lakhs i.e. @ 15% per annum, and profits have gone up from Rs. 390 lakhs (6%) to Rs. 1903 lakhs (17%).

NB: 10 Lakhs = 1 million

## Quality Indices:

A few abstract of quality indices compiled by the quality control & quality Assurance departments are enclosed.

## Task:

You are required to select 3 potential areas for VE/VM Studies, indicating the ranking and reasons for the same.

# QUALITY INDICES FOR IN COMING MATERIALS AND

# SUB-CONTRACT OPERTIONS FOR THE PAST TWO YEARS

| SI No | Index | Inward goods | | Sub- contact | | Total | |
|---|---|---|---|---|---|---|---|
| | | 1978 | 1979 | 1978 | 1979 | 1978 | 1979 |
| 1. | Value of goods inspected (Rs. in lakhs) | 3227 | 3249 | 80.5 | 79.4 | 2207.5 | 3328.4 |
| 2. | Values of goods rejected (Rs.in lakhs) | 128.7 | 128.0 | 2.2 | 3.36 | 130.9 | 131.36 |
| 3. | Percentage of rejection | 3.99 | 3.94 | 2.73 | 4.23 | 3.96 | 3.95 |
| 4. | Value of goods accepted with deviation (in lakhs) | 111.9 | 110.8 | 4.73 | 0.23 | 116.63 | 111.03 |
| 5. | Percentage of goods accepted with deviation | 3.46 | 3.41 | 5.87 | 0.29 | 3.52 | 3.36 |
| 6. | Value of goods accepted after re-work. (Rs. in lakhs) | 43.48 | 0.01 | 6.93 | 0.24 | 50.41 | 0.25 |
| 7. | % of goods accepted after rework | 1.35 | 0.00 | 8.61 | 0.30 | 1.52 | -- |
| 8. | No. of SRVs* recd | 13,000 | 10,654 | 4737 | 5416 | 17,746 | 16,070 |
| 9. | No. of SRVs cleared | 12,729 | 10,347 | 4737 | 5416 | 17,466 | 15,763 |
| 10. | % of SRVs Cleared | 97.8 | 97.12 | 100 | 100 | 98.4 | 98.09 |

SRV= Store Receipt Voucher also called Goods Receipt

# ANALYSIS OF NON-CONFORMANCES

### (Re-work, Rejections and Deviations)

|  | 1979 | 1980 |
|---|---|---|
| 1. Value of total outturn | 11,320 | 13,970 |
| (Rs. in lakhs) |  |  |
| 2. Value of rework & rejections | 1,821 | 1,809 |
| (Rs. in lakhs) |  |  |
| 3. Ratio (2) to (1) | 16,88% | 12,95% |

A. CAUSE-WISE DETAILS OF REWORK & REJECTIONS:

(Figures indicate nos. of quality attributes inspected)

| Sl No | Agency | REWORK | | | | REJECTIONS | | | |
|---|---|---|---|---|---|---|---|---|---|
|  |  | 1979 | | 1980 | | 1979 | | 1980 | |
|  |  | Nos. | % | Nos. | % | Nos. | % | Nos. | % |
| 1. | Manufacturing. | 2901 | 28.52 | 11637 | 43.60 | 9946 | 54.74 | 13640 | 48.63 |
| 2. | Stores & purchase | 2420 | 23.79 | 12190 | 45.70 | 3201 | 17.62 | 7692 | 27.43 |
| 3. | Castings & forgings | 675 | 6.64 | 457 | 1.7 | 1694 | 9.22 | 1361 | 4.85 |
| 4. | Design | 220 | 2.16 | 1452 | 5.44 | 1575 | 8.67 | 772 | 2.75 |
| 5. | Process/Techy. | 118 | 1.18 | 39 | 0.14 | 464 | 2.55 | 281 | 1.00 |
| 6. | Sub-contract coordination | 2298 | 22.59 | 706 | 2.65 | 376 | 2.07 | 2678 | 9.55 |
| 7. | Others (QC, M&S, etc.) | 1539 | 15.13 | 206 | 0.77 | 914 | 5.03 | 1623 | 5.79 |
|  | Total : | 10171 | 100% | 26687 | 100% | 18170 | 100% | 28047 | 100% |

B. DETAILS OF DEVIATIONS:

|  | 1979 | | | 1980 | |
|---|---|---|---|---|---|
|  | Nos. | % | | Nos. | % |
| 1. Accepted in "As is" state | 4288 | 57.90 | | 4738 | 67.18 |
| 2. Accepted with change in mating part. | 821 | 11.09 | | 1033 | 14.65 |
| 3. Accepted with Rework | 1726 | 23.30 | | 916 | 12.99 |
| 4. Rejected | 570 | 7.69 | | 366 | 5.19 |
| Total: | 7405 | 100% | | 7053 | 100% |

# QUALITY INDICES FOR G.I. CASTING FOR THE PAST 3 YEARS

A.  PRODUCTION & INSPECTION:

Wt. in tonnes

(Rounded off to nearest tonnes)

| SI No | | 1978 | | 1979 | | 1980 | |
|---|---|---|---|---|---|---|---|
| | | Wt. | % | Wt. | % | Wt. | % |
| 1. | Castings produced | 1323 | 100 | 1484 | 100 | 1148 | 100 |
| 2. | Castings inspected | 1248 | 94.33 | 1434 | 96.63 | 1081 | 94.16 |
| 3. | Castings rejected | 223 | 18.7 | 184 | 12.88 | 148 | 13.70 |
| 4. | Castings despatched | 1042 | -- | 1376 | -- | 1031 | -- |
| B. | CAUSE-WISE REJECTION: | | | | | | |
| 1. | Production | -- | -- | 8878 | -- | -- | -- |
| 2. | Technology | -- | -- | 6159 | -- | -- | -- |
| C. | DEFECT-WISE REJECTION | | | | | | |
| 1. | Mis-run | 6324 | 3.7 | 137 | 0.9 | 880 | 4.4 |
| 2. | Shift | 29656 | 17.4 | 1188 | 8.2 | 1459 | 7.3 |
| 3. | Mech. damage | 6138 | 3.6 | 4 | -- | 396 | 2.0 |
| 4. | Cold Shut | | | | | | |
| 5. | Blow holes | 15392 | 9.0 | 903 | 6.3 | 538 | 2.7 |
| 6. | Shrinkage | 26137 | 15.4 | 5011 | 34.5 | 7750 | 39.00 |
| 7. | Slag inclusions | 29934 | 17.6 | 2327 | 16.0 | 5044 | 25.4 |
| 8. | Sand inclusions | 29934 | 17.6 | 2327 | 16.0 | 5044 | 25,4 |
| 9. | Cracks | 9052 | 5.6 | 610 | 4.2 | 80 | 0.4 |
| 10. | Other defects | 19260 | 11.3 | 3919 | 26.9 | 2660 | 14.4 |
| | Total: | 170136 | | 14531 | | 19856 | 100 |

# INFORMATION/DATA COLLECTION

* **MARKETING/COMMERCIAL:**
    - Customer's requirements

      (Reliability, Serviceability, etc.)
    - Market demands.
    - Marketing strategies.

* **DESIGN/ENGINEERING:**
    - Dimensioned drawings

      (of parts and assemblies)
    - Latest specifications

      (fits, finish, tolerances, etc.)
    - Weight of parts.
    - Operating conditions.

      (Max. & Min. Temps., pressures and other environmental factors.)

* **PRODUCTION PLANNING & CONTROL:**
    - Scheduling.
    - Capacity (make or buy decision).
    - Batch size.

* **METHODS ENGINEERING:**
    - Process specifications

      (Machine tool, J&F, Heat and Surface treatment, etc)
    - Material preparation plans.

      (Material utilisation, cutting allowances, etc.)
    - Work standards/Operation times.
    - Level of skills.

* <u>MATERIALS MANAGEMENT:</u>

    – Suppliers (Current & Potentials)

    – Purchasing cost.

    – Vendors performance & development.

    – Delivery schedules.

    – Inventory carrying costs.

    – Max, and min, stocks and anticipated arrivals.

    – Storage, handling, distribution, disposal, etc.

    – Scrap records.

    – Cost of bought-out items.

* <u>PRODUCTION/MANUFACTURING:</u>

    – Process constraints.

    – 'Most likely' delivery date.

    – Deviation and re-work.

    – Available skills.

    – Models or prototype of parts or complete set of dismantled components.

* <u>QUALITY CONTROL & TESTING:</u>

    – Incidence of rejection & rework.

    – Third party quality requirements.

    – Inspection and testing procedures.

    – Limits of deviation allowed.

* <u>COSTING/FINANCE:</u>

    – Product cost in terms of materials, labour, overheads, etc.

    – Cost of rework & rejections.

    – Full analysis of work costs.

    – Value added/Conversion cost.

    – Return of Investment.

\* <u>PACKING AND DESPATCH:</u>

- Packaging

- Handling

- Transportation

- Ware-housing

- Distribution/Delivery of finished goods.

\* <u>MISCELLANEOUS:</u>

- Maintenance requirements

- Safety requirements

- In case of consumables;

- Performance data

- Consumption pattern

- Environmental influence

- Special care in storage, handling, disposal, shelf-life, etc.

(Ex. 1. Cutting oil causes dermatitis; hence it should be non-corrosive to job and machine tool, components, and non-toxic/non-irritant.

2. Some materials like cutting oils & lubricants should remain stable during service life, and be compatible with each other.

3. Dust of H.G.L. & F.G.R.P. and fumes of acids, paints and other chemicals are injurious to health).

# DEFINITION OF FUNCTION

## (A) EXAMPLES OF VERBS AND NOUNS:

|  | VERBS | NOUNS | |
|---|---|---|---|
|  |  | Measurable | Non-Measurable |
| WORK FUNCTIONS | Amplify | Air | Article |
|  | Attract | Current | Balance |
|  | Collect | Corrosion | Circuit |
|  | Conduct | Dimension | Damage |
|  | Change | Erosion | Device |
|  | Circulate | Flow | Direction |
|  | Dissipate | Heat | Equilibrium |
|  | Emit | Light | Part |
|  | Filter | Leverage | Repair |
|  | Hold | Power | Table |
|  | Indicate | Pressure | Support |
|  | Induce | Radiation |  |
|  | Illustrate | Surface |  |
|  | Insulate | Time |  |
|  | Prevent | Thrust |  |
|  | Protect | Torque |  |
|  | Provide | Velocity |  |
|  | Position | Weight |  |
|  | Transport |  |  |
|  | Transmit |  |  |
| SELL FUNCTIONS | Attract |  | Appearance |
|  | Create |  | Beauty |
|  | Decrease |  | Convenience |
|  | Enhance |  | Cost |
|  | Improve |  | Design |
|  | Increase |  | Effect |
|  | Make |  | Feature |
|  | Obtain |  | Form |
|  | Indicate |  | Impact |
|  |  |  | Looks |
|  |  |  | Prestige |
|  |  |  | Reputation |
|  |  |  | Style |

231

(B) <u>USE OF VERBS & NOUNS:</u>

| ITEMS (What is it?) | FUNCTION (What does it do?) |
|---|---|
| 1. Screw or rivet | Fasten parts |
| 2. Lamp | Give light |
| 3 Spanner | Provide leverage |
| 4. Paint | protest surface |
| 5. Lock washer | resist rotation |
| 6. Engine | Provide motive power |
| 7. Steering wheel | facilitate manoeuvring |
| 8. Bonnet | Protect engines |
| 9. Elec. Horn | emits sound |
| 10. Speedometer | indicate speed |
| 11. Valve | control flow |
| 12. Pen | make marks |
| 13. Heat sink | dissipate heat |
| 14. Watch | Indicate time |
| 15. Compact Disc | record data |
| 16. Bell bearing | reduce friction |
| 17. Exposed brick facia | enhance beauty or looks |
| 18. Crane hook | hold sling loops |
| 19. Silica gel | absorb moisture |
| 20. Spectacles | correct vision |

## FUNCTIONAL ANALYSIS
## PRODUCT : INK POT

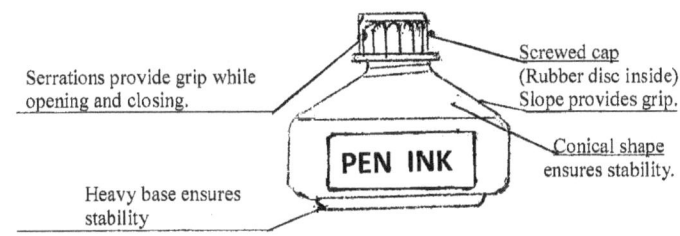

Serrations provide grip while opening and closing.

Screwed cap (Rubber disc inside)
Slope provides grip.

Conical shape ensures stability.

Heavy base ensures stability

PEN INK

| SI No | DESCRIPTION OF ASSEMBLY, PART OR ELEMENTS | Cost | | | | | FUNCTION PRIMARY | | FUNCTION SECONDARY | | | Total Cost. |
|---|---|---|---|---|---|---|---|---|---|---|---|---|
| | | Material | Labour | Over head | Others (tooling) | Total | Hold Ink | Provide stability | Prevent Leakage | Provide grip | Facilitate identification | |
| 1 | Bottle & its shape | | | | | | | | | | | |
| 2 | Cap & its shape | | | | | | | | | | | |
| 3 | Rubber disc and punching out. | | | | | | | | | | | |
| 4 | Label printing | | | | | | | | | | | |
| 5 | Assembly | | | | | | | | | | | |
| | - Fit disc in cap | | | | | | | | | | | |
| | - Paste label | | | | | | | | | | | |
| | - Fill ink. | | | | | | | | | | | |
| | Cost of Function. | | | | | | | | | | | |
| | Worth of function. | | | | | | | | | | | |

## FUNCTIONAL COST ANALYSIS

## PRODUCT : 2 PIN PLUG FOR ELECTRICAL CONNECTIONS

| Sl No | Description of sub – assembly, component/part and feature. | COST | | | | | | FUNCTIONS | | | | | | | | |
|---|---|---|---|---|---|---|---|---|---|---|---|---|---|---|---|---|
| | | | | | | | | PRIMARY | | | | SECONDARY | | | TER-TIARY | Total |
| | | Material | Labour Machining | Over-head | Others | Total | Make Contact | Facilitate Cable Connection | Provide Safety | Provide Tension | Total Resist Movement | Prevent Corrosion | Provide Grip | Assemble parts | |
| 1 | 2 | 3 | 4 | 5 | 6 | 7 | 8 | 9 | 10 | 11 | 12 | 13 | 14 | 15 | 16 |
| 1 | Cylindrical Pins (2 Nos.) with slits at one end and threads at the other end. | | | | | | | | | | | | | | |
| 2 | Square connectors (2 Nos.) each with internal threads at both ends and a slotted head screw. | | | | | | | | | | | | | | |
| 3 | Base with two holes. | | | | | | | | | | | | | | |
| | – a separator and threads on periphery (bought out) | | | | | | | | | | | | | | |
| 4 | Threaded Cap with serration for finger grip. (bought out) | | | | | | | | | | | | | | |
| 5 | Assembly (labour) | | | | | | | | | | | | | | |
| 6 | Function Cost : Total : | | | | | | | | | | | | | | |
| 7 | Worth | | | | | | | | | | | | | | |

234

# FUNCTIONAL COST ANALYSIS

(Trainer's Copy)

## PRODUCT : 2 – PIN PLUG FOR ELECTRICAL CONNECTIONS

(Figures in RUPEES)

VA-08-A/VI

| SI No | Description of sub – assembly, component/part and feature. | COST | | | | | FUNCTIONS | | | | | | | | Total |
|---|---|---|---|---|---|---|---|---|---|---|---|---|---|---|---|
| | | | Labour | | | | PRIMARY | | | | SECONDARY | | | TER-TIARY | |
| | | Material | Machining | Over-head | Others | Total | Make Contact | Facilitate Cable connection | Provide Safety | Provide Tension | Total Resist Movement | Prevent Corrosion | Provide Grip | Assemble parts | Total |
| 1 | 2 | 3 | 4 | 5 | 6 | 7 | 8 | 9 | 10 | 11 | 12 | 13 | 14 | 15 | 16 |
| 1 | Cylindrical Pins (2 Nos.) with slits at one end and threads at the other end. | 3.0 | 0.8 | 1.2 | 0.2 | 5.2 | 1.0 | 0.9 | -- | 1,3 | -- | 2,0 | -- | -- | 5,2 |
| 2 | Square connectors (2 Nos.) each with internal threads at both ends and a slotted head screw. | 0.85 | 0.25 | .45 | .50 | 1.6 | 0.4 | 0.4 | -- | -- | 0.4 | 0.4 | -- | -- | 1.6 |
| 3 | Base with two holes. – a separator and threads on periphery (bought out) | -- | -- | -- | -- | 0.8 | -- | -- | 0.6 | -- | -- | -- | 0.2 | | 0.8 |
| 4 | Threaded Cap with serration for finger grip. (bought out) | -- | -- | -- | -- | 1.1 | -- | -- | 0.7 | -- | -- | -- | 0,2 | 0.2 | 1.1 |
| 5 | Assembly (labour) | -- | 12 | 17 | 1 | 30 | -- | -- | -- | -- | -- | -- | -- | 30 | 30. |
| 6 | Function Cost: Total : | -- | -- | -- | -- | -- | 14 | 13 | 13 | 13 | 4 | 24 | 2 | 7 | 90 |
| 7 | Worth ** | | | | | | | | | | | | | | |

** To be filled in by 'consensus' during the course

235

# FUNCTION LEVELS

LEVEL 1

LEVEL 2

LEVEL 3

LEVEL 4

Functions of the product

Functions of Sub-Assemblies to achieve the functions of the product

Functions of the components to achieve the functions of Sub-Assly. to achieve the functions of the product

Functions of each feature of component to achieve the function of the component to achieve the function of Sub-Assly. to achieve the function of the product.

F1  F2  F3  F4  F  F5  F6  F7  F8  F9  F10  F11  F12  F13  F14  F

236

# VALUE ENGINEERING  CASE STUDY-2

## CORE OF PAPER REEL

1.    Introduction and project Selection:

M/s Mysore Paper Mills Ltd., Bhadravati, manufacture about 75,000 tonnes of paper wound on 10′ long hollow cores, which are later on cut into 2/3 parts each, for end use. The winding, transportation and end use call for a rigid hollow core, which can be cut by the same knives as are used for cutting the paper reels. The annual consumption is about 50,000 cores.

Initially such cores were obtained from a local source by supplying the raw materials (here, craft paper) and paying for conversion into cores. As the production picked up, the requirement of reels went up but the market demand for craft paper outstripped production capacity, resulting in acute scarcity of craft paper for manufacture of the reels.

Hence, to meet the production demand of reels, it was decided to dispense with the craft paper cores and use alternative materials. Also as a cost reduction idea, straw board tubes were ordered and used; but in stacking and transportation, heavy damages were incurred, resulting in customer dissatisfaction and the paper reels returning to "the first hopper" in the mill.

At this stage, the decision was taken to go back to craft paper tubes but the original conventor had wound up his business and new supplier quoted nearly double the original rates (using their raw materials). The difference in rate amounts to an additional cash outflow of Rs.47 lakhs per year at present rates of production.

2.    The use of VA/VE Approach:

At this juncture, the VA/VE team approach was adopted to look for value opportunities and the outcome is quite encouraging.

All the steps in VA/VE methodology have been practiced as under:

2.1   Information Phase:

Information was gathered from the reliable sources as follows:

(a)   Design Department:
- Dimensioned drawing;
- Material Specifications;
- Technical parameters like rigidity, behaviour under different environment, etc.

(b)   Material Department:
- Vendors, their background, process capability, etc.:
- Price; Lead-Time;
- Transportation Cost;
- Minimum & Maximum Levels of Inventory;
- Inventory carrying cost;
- Availability of raw-materials, handling methods, etc.

(c)   Production:
- In process requirement as regards to bore diameters, surface roughness, etc.;

(d)   Marketing:
- Customers' preferences;
- Dimensional requirements;
- Re-use possibilities/end uses, etc.

The above information is given in Appendix-I.

2.2   Function – Cost Analysis:

(a)   The Basic Function of the Reel Core was identified by the team as:

- "Carry Paper"

The Secondary Functions were defined as:

- Hold paper;
- Facilitate Winding;
- Facilitate Re-winding;
- Facilitate Transportation

(b)   The factor/attributes to meet the above functional requirements were discussed and listed as follows:

- Rigidity;
- Machineability;

- Weight;
- Surface finish;
- Cost;
- Availability;
- Lead time.

Since there is only one basic function, the total cost is assumed, by the team, to be attributable mostly to that function. The worth of the basic function was arrived at, by comparison with other items performing the same function and it was estimated at Rs.80/- per core as against the existing cost of Rs.195/-, value ratio = $\frac{80}{195}$ = 0.4 showed a tremendous value opportunity.

Synopsis of this function cost analysis is shown in Appendix-I.

2.3   Creative phase/Speculating Alternatives:

The team conducted a Brainstorming Session on alternatives to 'carry paper' and produced 27 ideas, ranging from the simplest to space-age variety. The ideas are listed in Appendix-II.

2.4   Evaluation phase;

After an overnight incubation, when the team met for evaluation, 6 more ideas were generated, which are also included in the Appendix-II.

Using the most essential attributes as the first "sieve," 27 ideas were dropped like 'hot potatoes' and the short-listed 7 were subjected to close scrutiny, using Forced Decision Matrix (weighted value – T chart). (Please refer to appendixes III & IV).

The 3 prospective 'winners' are then ranked and earmarked for further investigation on priority. It does not mean that the rest are rejected, but due to very obvious reason like indigenous availability, lack of knowledge on process capabilities etc., they are either "held back" or given second priority for investigation.

2.5   Recommendation Phase:

The following three alternatives are recommended for trial run/placing trial orders on suppliers:

| Order of Priority | Alternative | Approx. cost of development |
|---|---|---|
| | | Rs. |
| 1. | Straw board mixed with grey mill board cores. | 10,000-00 |
| 2. | Corrugated card board tubes. | 15,000-00 |
| 3. | Forming (in-situ) core in Process of winding | 5,000-00 |

3.  Conclusion:

The cost of developing any of the above three alternatives is negligible compared to the annual costs of reel cores.

Granting that all the three processes may be developed simultaneously, the total development cost can be recovered In one day's consumption of reel cores. Considering envisaged annual VA gains of the order Rs.4.0 to Rs. 5.0 millions this project needs to be expedited for conclusive trial runs and implementation.

## INFORMATION/DATA

### Kraft Core Reel

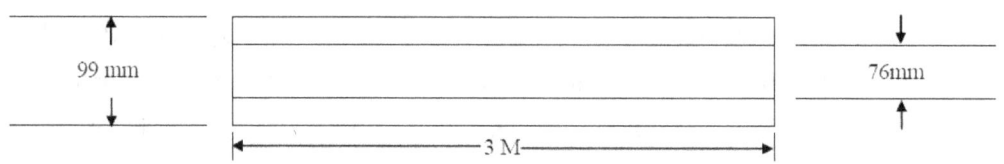

Specs:      Raw material  :      60 GSM Kraft paper;

          Adhesive      :      Glue

Load carrying cap:    250 kgs static and 1000 kgs Impact

Function – Cost Analysis

| Item | Basic Function | | Present unit cost Rs. | Functional worth Rs. | Value index |
|------|------|------|------|------|------|
| | Verb | Noun | | | |
| Kraft paper Core | Carry | Paper | 195 | 80 | 0.4 |

Annual requirement  :      50,000 pieces

Annual cost         :      Rs. 97,50,000/-

Ideas listing on: "what else will 'carry paper'?

| S.No. | Ideas |
|-------|-------|
| 1. | Wooden Cylindrical tube |
| 2. | Seamless tube |
| 3. | PVC pipe |
| 4. | FRP pipe |
| 5. | SMC pipe |
| 6. | MS fab/welded pipe |
| 7. | Wire mesh pipe welded |
| 8. | Re-used core pipes |
| * 9. | Use corrugated pipes |
| 10. | Circular disc with |
| * 11. | Form pipe in process of paper |
| * 12. | Structural foam pipe |
| 13. | Coil springs |
| 14. | A.C.C. pipe |
| 15. | Plywood pipe |
| * 16. | Polypropylene pipe |
| * 17. | Straw board pipe |
| * 18. | Mill board pipe |
| 19. | Cork pipe |
| 20. | Bamboo pipe |
| 21. | Honey comb pipe |
| 22. | Thermocol pipe |
| 23. | Bubble core |
| 24. | X – shape wooden reapers |
| 25. | Stack paper in flats |
| 26. | Paper mache tubes |
| 27. | Tube spleened in small lengths |
| 28. | Aluminium Tubes |
| 29. | Earthen pipe |
| 30. | Spun pipe |
| 31. | Sintered pipe |
| 32. | Hylam pipe |
| 33. | Rubberised pipe |

_____

* Ideas short-listed for further evaluation.

A) WEIGHTING OF ATTRIBUTS :

| Sl No | Attribute | Code | Weightage |
|---|---|---|---|
| 1. | Rigidity | A | 0.25 |
| 2. | Machinability | B | 0.10 |
| 3. | Weight | C | 0.05 |
| 4. | Surface Finish | D | 0.05 |
| 5. | Cost | E | 0.25 |
| 6. | Availability | F | 0.25 |
| 7. | Lead Time | G | 0.05 |
| Total for all attributes | | | 1.00 |

B) FUNCTIONAL RATTING MATRIX :

| Sl No | Alternatives | A | B | C | D | E | F | G |
|---|---|---|---|---|---|---|---|---|
| 1. | Straw/Mill board | 0.70 | 0.80 | 0.90 | 0.40 | 1.00 | 1.00 | 0.80 |
| 2. | Structural Plastic | 1.00 | 0.40 | 0.70 | 0.90 | 0.40 | 0.40 | 0.40 |
| 3. | Polypropylene | 0.70 | 0.40 | 0.40 | 0.80 | 0.50 | 0.50 | 0.40 |
| 4. | Corrugated board | 0.80 | 1.00 | 1.00 | 1.00 | 0.70 | 0.80 | 0.80 |
| 5. | Forming core in process | 0.80 | 0.80 | 0.80 | 1.00 | 0.60 | 1.00 | 1.00 |
| 6. | Paper Mache | 0.90 | 0.80 | 0.80 | 0.90 | 0.70 | 0.80 | 0.80 |
| 7. | Kraft paper (existing) | 0.80 | 0.80 | 0.90 | 1.00 | 0.60 | 080 | 0.80 |

# FEASIBILITY  RANKING  MATRIX

| SI No | Alternatives | A | B | C | D | E | F | G | Total Points | Rank |
|---|---|---|---|---|---|---|---|---|---|---|
| 1. | Straw/mill board | 0.175 | 0.090 | 0.020 | 0.050 | 0.250 | 0.250 | 0.040 | 0.875 | I * |
| 2. | Structural plastic | 0.250 | 0.070 | 0.045 | 0.020 | 0.100 | 0.100 | 0.020 | 0.605 | |
| 3. | Poly propylene | 0.175 | 0.040 | 0.040 | 0.025 | 0.125 | 0.125 | 0.020 | 0.550 | |
| 4. | Corrugated Board | 0.200 | 0.100 | 0.050 | 0.035 | 0.175 | 0.200 | 0.040 | 0.800 | II * |
| 5. | Forming Core in process | 0.175 | 0.080 | 0.050 | 0.050 | 0.150 | 0.250 | 0.050 | 0.785 | III * |
| 6. | Paper Mache | 0.125 | 0.080 | 0.045 | 0.035 | 0.175 | 0.200 | 0.040 | 0.700 | |
| 7. | Kraft paper (existing) | 0.175 | 0.090 | 0.050 | 0.025 | 0.150 | 0.200 | 0.040 | 0.730 | |

*   Recommended for conclusive trial runs and adoption.

# HAND-OUTS FOR GENERAL READING

## ON VE/VM AND CREATIVITY

| S. No. | Hand out No./ Session No. | Description/Source/Author |
|---|---|---|
| 1. | VM-01/II | Techniques of Value Engineering. (VE Guides) (Lawrence Miles) |
| 2. | VN-02/II | The twelve question – Framework of VA/VE. |
| 3. | VM-03/VI | Function – Level of study (Functional levels of a Transport Vehicle) |
| 4. | VM-04/VII | The Calf Path – poem (Sam Walter Foss) |
| 5. | VM-05/VII | It couldn't be done – a poem (Edger A. Guest) |
| 6. | VM-06/VII | Sixty – three ways to stop creativity (Management of Intelligence – Gregory) |
| 7. | VM-07/VII | As some men see us – a poem |
| 8. | VM-08/VII | A definition of a design (Leo Tse) |
| 9. | VM-09/XI | Little Red Hen – a poem. |

NB: The Trainer/Facilitator may add some more relevant material to the above list.

## TECHNIQUES OF VALUE ENGINEERING (V.E. GUIDES)

Lawrence Miles lists 13 as follows:

1.  Avoid generalities (Be Specific)

2.  Get maximum cost data.

3.  Use information from only the best source.

5.  Use real creativity.

6.  Acknowledge, identify and overcome road blocks.

7.  Use Industry specialists to extend specified knowledge.

8.  Utilise suppliers' available functional products.

9.  Utilise and pay for suppliers' skills and knowledge.

10. Utilise speciality processes.

11. Highlight the cost of key tolerances and finishes.

12. Use applicable standard parts.

13. Use criterion "Would I spend my money this way"

*************

# THE TWELVE QUESTIONS -FARMEWORK OF V.A/V.E

| **QUESTIONS** | **FOR JOB PLAN PHASE** |
|---|---|
| 1. What is it? | Orientation and Information |
| 2. What does it cost? | |
| 3. How many parts? | |
| 4. What does it do? | |
| 5. How many required? | |
| 6. Which is primary function? | |
| 7. What else will do? | Speculation |
| 8. What will that cost? | Evaluation |
| 9. Which are cheapest? (Which three of the alternative ways show the greatest difference between Cost and Use Value?) | |
| 10. Which ideas are to be developed? | Program planning & Execution |
| 11. What other functions (Work or Sell) | |
| 12. What do we need to sell our ideas And forestall road blocks? | |

——————xxx——————

R.G. CHAUDHARI

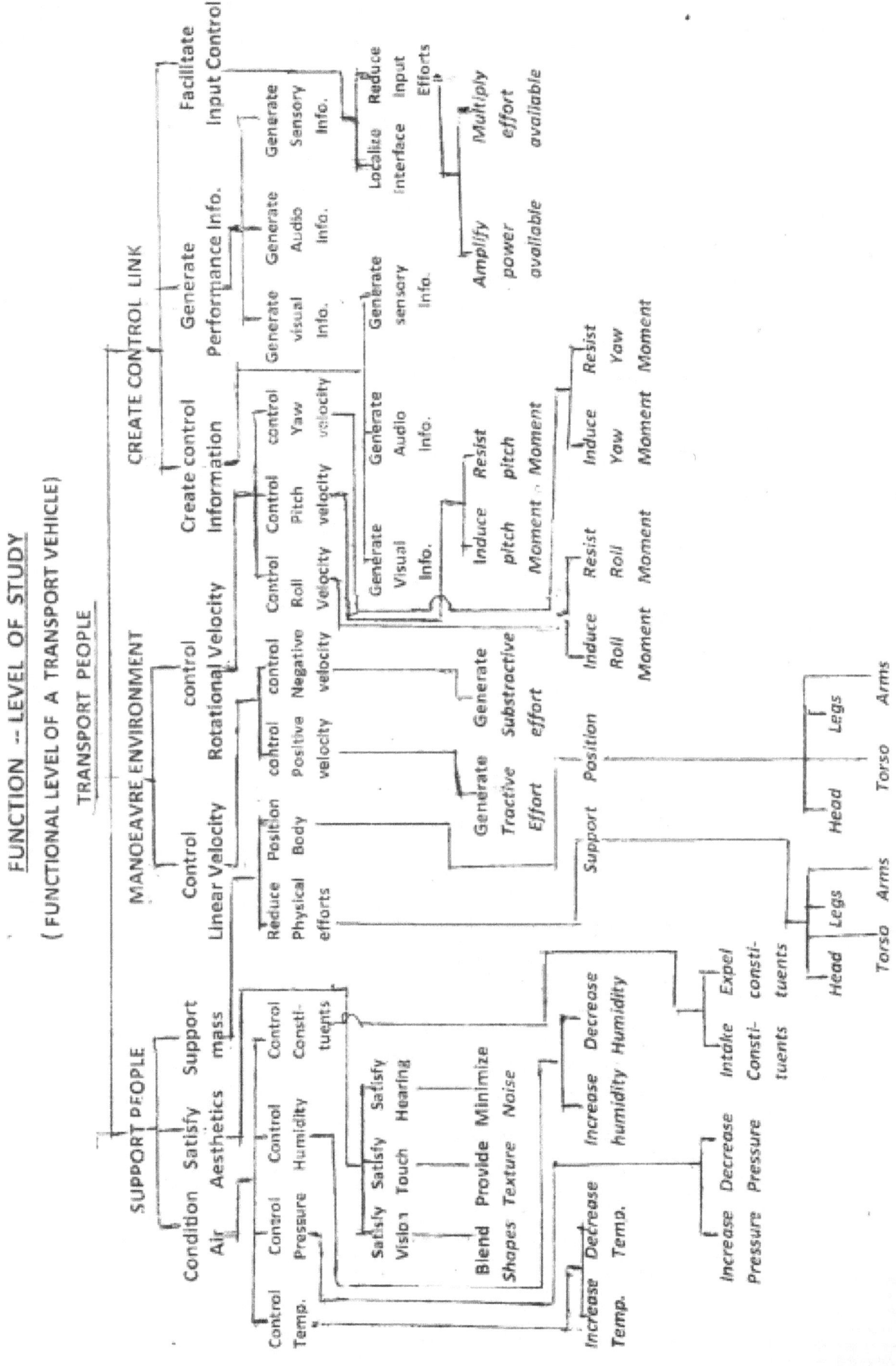

249

### *THE CALF PATH*

-Sam Walter Foss.

*One day through the primeval wood,*
*a calf walked home as good calves should;*
*But made a trail all bent askew, a*
*crooked trail as all calves do.*
*Since then three hundred years have field,*
*and I infer the calf is dead.*
*But still he left behind his trail,*
*and thereby hange my moral tale.*
*The trail was taken up next day,*
*by a long dog that passed that way;*
*And then a wise bell-weather sheep,*
*And draw the flock behind him, too;*
*as good bell-weathers always do.*
*And from that day, o'er hill and glade,*
*as good bell-weather always do.*
*And from that day, o'er hill and glade,*
*through those old woods a path was made,*
*And many men wound in and out,*
*and dodged and turned and bent about,*
*And uttered words of righteous wrath,*
*because it was such a crooked path;*
*But still they followed – do not laugh*
*the first migrations of that calf*
*And through this winding wood-way stalked,*
*because he webbled when he walked,*
*This forest path became a lane,*
*that bent and turned and turned again;*
*This crooked lane became a road,*
*Where many a poor horse with his load,*
*Toiled on beneath the burning sun,*
*and travelled some three mile in one.*
*And thus a century and a half,*
*They trod the footsteps of that calf.*

....2/-

*The years passed on in swiftness fleet,*

*The road became a village street,*

*And this, before men were aware,*

*a city's crowded throughfare.*

*And soon the central street was this,*

*of renowned metropolis;*

*And men two centuries and half,*

*trod in the footsteps of that calf.*

*Each day a hundred thousand men,*

*follow this zigzag calf again,*

*And o'er his crooked journey went,*

*the traffic of a continent,*

*A hundred thousand men were led,*

*by one calf near three centuries dead.*

*They followed still his crooked way,*

*and lost one hundred years a day.*

*For thus such reverence is lent,*

*to a well-established precedent.*

*A moral lesson this might teach,*

*were I ordained and called to preach;*

*For men are prone to go it blind,*

*along the calf-path of the mind,*

*And work away from sun to sun,*

*to do what other men have done.*

*They follow in the beaten truck,*

*and in and out, and forth and back,*

*And still their devious course pursue,*

*to keep the path that others do.*

*They keep the path a sacred groove,*

*along which all their lives they move;*

*But how the verse old wood-gods laugh,*

*who first saw the primeval calf.*

*Ah, many things this tale might teach*

*but I am not ordained to preach.*

## IT COULDN'T BE DONE
### (Edgar A Guest)

Somebody said that it couldn't be done.

But he with a chuckle replied

That 'may be it couldn't but he would be one

Who wouldn't say to till he tried.

So he buckled right in with the trace of a grain

On his face. If he worried he hid it

He started to sing as he tackled the thing

That couldn't be done, and he did it.


Somebody scoffed "you'll never do that"

At least no one has ever done it.

But he took off his coat and he took off his hat,

And the first thing we knew he'd begun it.

With a lift of his chin and a bit of a grin

Without any doubting or quid it;

He started to sing as he tackled the thing

That couldn't be done, and he did it.


There are thousands to tell you it cannot be done.

There are thousands to prophesy failure,

There are thousands to point out to you one by one,

The dangers that wait to assail you.

But just buckle in with a bit of a grin,

Just take off your coat and go to it:

Just start in to sing as you tackle the thing

That "cannot be done," and you will do it.

—--xxx—--

# SIXTY THREE WAYS TO STOP CREATIVITY

## (Source: Management of Intelligence: Gregory)

A good idea But….

Against company policy

Ahead of the times

All right in theory

Be practical

Can you put it into practice?

Costs too much.

Don't start anything yet.

Have you considered….

I know it won't work

It can't work

It doesn't fit human nature.

It has been done before

It needs more study.

It's not budgeted

It is not good enough.

It is not a part of your job

Let me add to that…

Let's discuss it

Let's form a committee.

Let's make a survey first.

Let's not step on toes.

Let us put it off for a while.

Let us sit on it for a while.

Let's think it over for a while.

Not ready for it yet

Of course it won't work.

Our plan is different.

Some other time.

Surely know better.

That is not our problem.

The boss won't go for it.

It  will create industrial relation problems

The new men won't understand

The old timers won't use it.

The timing is off.

The Union won't go for it.

There are better ways.

They won't go for it.

Too academic.

Too hard to administer.

Too hard to implement.

Too late.

Too many projects now

Too much paper work.

Too old fashioned

Too Soon.

We have been doing it this way

For a long time and it works

We haven't the man power

We haven't the time.

We're too big.

We're too small.

We have never done it that way.

We've tried it before

What bubblehead thought that up?

What will the customer think..?

What will the Union think?

What your really saying is…

What do you think you are?

Who also has tried it?

Why hasn't someone suggested it

before if it's a good idea?

Your are off base

Don't you have better things to do?

—xxx—

## AS SOME MEN SEE US – A POEM
## -PRODUCT DESIGN ENGINEER.

The Designer bent across his board,

Wonderful things in his head were stored

And he said as he rubbed his throbbing bean

"How can make this thing tough to machine?

If his part were here only straight

I'am sure the thing would work first rate,

But it would be so easy to turn and bore

It would never make the machinists sore,

I better put in a right angle there

Then watch those babies tear their hair

Now I'll put the holes that hold the cap

Way down in where they're hard to tap.

Now this piece won't work I'll bet a buck

For it can't be held in shoe or chuck

It can't be drilled or it can't be ground

In fact the design is exceedingly sound".

He looked again and cried "At last-

Success is mine, it can't even be cast".

**********

## A DEFINITION OF A DESIGN

The wheel's hub holds thirty spokes Utility

depends on the hole through the hub.

The potter's clay forms a vessel.


A house is built with solid walls

The nothingness of window and door alone

renders it usable,

That which exists may be transformed

What is non-existent has boundless uses.

-Lao Tse

# LITTLE RED HEN

*No one really knows who wrote this updated version of the well-known fable. But it has been widely reprinted and even read at shareholders' meetings.*

Once upon a time, there was a little red hen who scratched about the barnyard until she uncovered some grains of wheat. She called her neighbors and said, "If we plant this wheat, we shall have bread to eat. Who will help me plant it?"

"Not I," said the cow.

"Not I" said the duck.

"Not I," said the pig.

"Not I," said the goose.

"Then I will," said the little red hen, and she did.

The wheat grew tall and ripened into golden grain.

"Who will help me reap my wheat?" asked the little red hen.

"Not I," said the duck.

"Out of my classification," said the pig.

"I'd lose my seniority," said the cow.

"I'd lose my unemployment compensation," said the goose.

"Then I will," said the little red hen, and she did.

At last it came time to bake the bread. "Who will help me bake the bread?" asked the little red hen.

"That would be overtime for me," said the cow.

"I'd lose my welfare benefits," said the duck.

"I'm a dropout and never learned how." said the pig.

"If I'm to be the only helper, that's discrimination," said the goose.

"Then I will," said the little red hen. She baked five leaves and held them up for her neighbors to see.

They all wanted some – in fact, demanded a share.

but the little red hen said, "No, I can eat the five loaves myself."

"Excess profits!" yelled the cow.

"Capitalist leech!" cried the duck.

"I demand equal rights!" shouted the goose

The pig just grunted. Then they hurriedly painted "unfair" picket signs and marched around, shouting obscenities.

The government agent came and said to the little red hen, "You must not be greedy".

"But I earned the bread," said the little red hen.

"Exactly," said the agent. "That is the wonderful free-enterprise system. Anyone in the barnyard can earn as much as he wants. But, under government regulations, the productive workers must divide their product with the idle."

And they lived happily ever after. But the little red hen's neighbours wondered why she never baked bread again.

## TEAM GAMES

### GAME-1: WIN AS MUCH AS YOU CAN.

The Trainer is advised to search On Google 'Win As Much As you Can'. Select www.trainingcoursematerial.com and other references./sites. Study them carefully and then play the game.

This game takes about 1 hour.

### GAME-2 : GAME OF 64 Cubes:

Search for manufacturers of plastic cubes on Google. Buy online one pack of 100 cubes of size 1 Cm. Issue 64 cubes to one team of 4 or 5 members. Ask the team to assemble as many regular geometrical figures as they can. Watch how one of them pulls max. cubes towards him.

Have all members participated? Draw conclusion on team work.

### GAME-3: GAME Of Balloons:

This is the easiest game requiring about 15 minutes.

- Select a team of about 10 members standing in a row.

- Give them balloons filled with air and tied with thread about a metre long.

- Announce that THE ONE LEFT WITH BALOON IS THE WINNER.

(Never use the word 'burst')

- Watch how they try to burst as many balloons as possible-

- When one is left with balloon, make them stand in a row and watch how the winner feels elevated.

- Now announce **that you never asked them to burst balloons**,

- Was it not possible for all of you to be the winners?

- Our psychology is that for winning, we should defeat others or put them to loss

- Had you not burst the balloons, every member of the team and the team would have been the winner

- This is what happens in families, Companies, Political Parties, Nations, etc.,

- In order to defeat others attacking individuals, groups, parties, etc, get themselves defeated/destroyed.

— xxx —

## DESIGN OF A PEN HOLDER

Normally, a house-wife keeps a ball pen or pencil in a kitchen shelf cluttered with other articles like cans, bottles, utensils, etc., and beyond the reach of small children, Consequently and invariably, the pen gets 'lost' or misplaced and it takes efforts to trace it on the shelf.

A group of Technician apprentices fresh from polytechnic were asked to help the lady by suggesting some means to keep the pen in pre-assigned place. The only condition imposed was that the pen should not be tied by thread and hung from a nail, since it is required at many places in the kitchen.

Problem : By consensus, the problem was worded as "Design a Pen Holder" suitable for a single ball pen (with Clip).

Function : "Hold pen"      (primary)

Ensure stability

Provide safety (secondary)

Suggested

alternatives : Out of 41 ideas generated in 10 minutes, 50% were variations of conventional pen stands of various designs & materials usually found on the tops of office tables, and the remaining 50% had some element of novelty. These suggested alternatives are illustrated in the enclosure ( Page 258).

Your tasks : 1. Prepare a list of factors or criteria for evaluating relative merits of the alternatives.

2. Subject the first five ideas to Criteria T-Chart technique and give your decision to 'accept' or 'reject' or 'hold on'.

3. Take the next set of 5 ideas and arrive at a decision after quantifying the criteria expressed in absolute or percentage form.

# PEN HOLDER

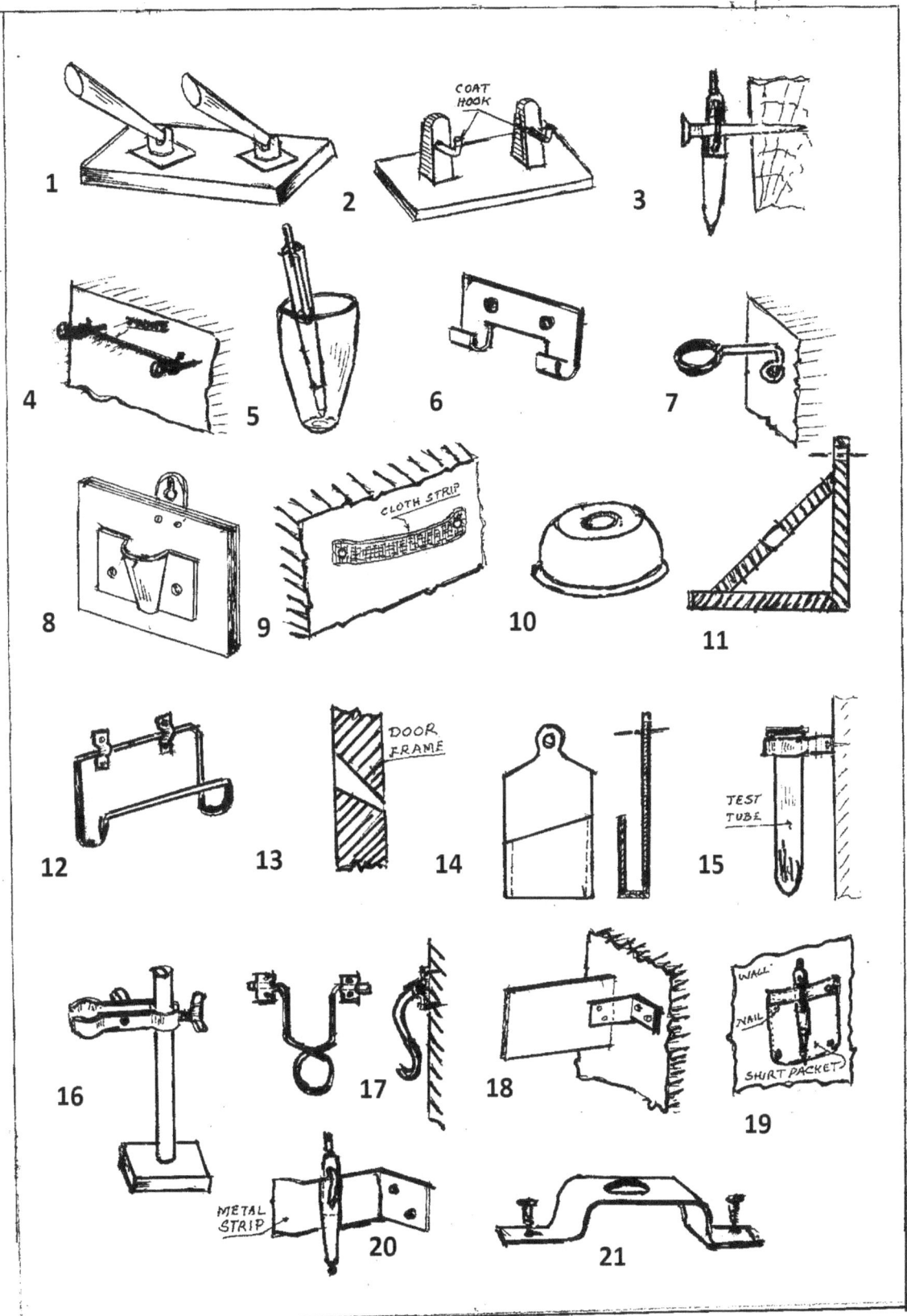

| | | |
|---|---|---|
| 30 | Bibliography of Value Technology – BY SAVE | " |
| 31. | Practical Techniques for Quick Reduction of Cost Work Design. -By QR Creasy. | " |
| 32. | FAST (Functional Analysis System Technique) Manual – Rand Creasy. | " |
| 33. | Techniques of Producing Ideas. - James W.Young. | British Publication. |
| 34. | Art of Thought – Wallas | " |
| 35. | How to Think up | Alex. F. Osborn. |
| 36. | Your Creative Power | " |
| 37 | Wake up your Mind | " |
| 38. | Applied Imagination | " |
| 39. | Investing in Value. D. Warburton Brown. | Asian Productivity Organisation. |
| 40 | Product Innovation – Knut Holt | Newnes – Butter Worths Mgt. Library. |

**NB: For further reading / study and advances, go to Internet and search under 'Value Engineering'.**

OOOOOOOOOOOOO